C0-DXD-321

Neocera, Inc.
10000 Virginia Manor Road, Suite 300
Beltsville, Maryland 20705-4215

Springer Series in **Materials Science** 28

Edited by R. M. Osgood

Springer Series in *Materials Science*

Advisors: M. S. Dresselhaus · H. Kamimura · K. A. Müller
Editors: U. Gonser · R. M. Osgood · M. B. Panish · H. Sakaki
Managing Editor: H. K. V. Lotsch

1 **Chemical Processing with Lasers**
By D. Bäuerle

2 **Laser-Beam Interactions with Materials**
Physical Principles and Applications
By M. von Allmen

3 **Laser Processing of Thin Films and Microstructures**
Oxidation, Deposition and Etching of Insulators
By. I. W. Boyd

4 **Microclusters**
Editors: S. Sugano, Y. Nishina, and S. Ohnishi

5 **Graphite Fibers and Filaments**
By M. S. Dresselhaus, G. Dresselhaus, K. Sugihara, I. L. Spain, and H. A. Goldberg

6 **Elemental and Molecular Clusters**
Editors: G. Benedek, T. P. Martin, and G. Pacchioni

7 **Molecular Beam Epitaxy**
Fundamentals and Current Status
By M. A. Herman and H. Sitter

8 **Physical Chemistry of, in and on Silicon**
By G. F. Cerofolini and L. Meda

9 **Tritium and Helium-3 in Metals**
By R. Lässer

10 **Computer Simulation of Ion-Solid Interactions**
By W. Eckstein

11 **Mechanisms of High Temperature Superconductivity**
Editors: H. Kamimura and A. Oshiyama

12 **Dislocation Dynamics and Plasticity**
By T. Suzuki, S. Takeuchi, and H. Yoshinaga

13 **Semiconductor Silicon**
Materials Science and Technology
Editors: G. Harbeke and M. J. Schulz

14 **Graphite Intercalation Compounds I**
Structure and Dynamics
Editors: H. Zabel and S. A. Solin

15 **Crystal Chemistry of High-T_c Superconducting Copper Oxides**
By B. Raveau, C. Michel, M. Hervieu, and D. Groult

16 **Hydrogen in Semiconductors**
By S. J. Pearton, M. Stavola, and J. W. Corbett

17 **Ordering at Surfaces and Interfaces**
Editors: A. Yoshimori, T. Shinjo, and H. Watanabe

18 **Graphite Intercalation Compounds II**
Editors: S. A. Solin and H. Zabel

19 **Laser-Assisted Microtechnology**
By S. M. Metev and V. P. Veiko

20 **Microcluster Physics**
By S. Sugano

21 **The Metal-Hydrogen System**
By Y. Fukai

22 **Ion Implantation in Diamond, Graphite and Related Materials**
By M. S. Dresselhaus and R. Kalish

23 **The Real Structure of High-T_c Superconductors**
Editor: V. Sh. Shekhtman

24 **Metal Impurities in Silicon Device Fabrication**
By K. Graff

25 **Optical Properties of Metal Clusters**
By U. Kreibig and M. Vollmer

26 **Gas Source Molecular Beam Epitaxy**
Growth and Properties of Phosphorus Containing III-V Heterostructures
By M. B. Panish and H. Temki

27 **Physics of New Materials**
Editor: F. E. Fujita

28 **Laser Ablation**
Principles and Applications
Editor: J. C. Miller

John C. Miller (Ed.)

Laser Ablation

Principles and Applications

With 85 Figures

Springer-Verlag
Berlin Heidelberg New York
London Paris Tokyo
Hong Kong Barcelona
Budapest

Dr. John C. Miller
Chemical Physics Section, Health Sciences Research Division
Oak Ridge National Laboratory, Building 4500S
Mail Stop 6125, Oak Ridge, TN 37831-6125, USA

Series Editors:

Prof. Dr. U. Gonser
Fachbereich 12.1, Gebäude 22/6
Werkstoffwissenschaften
Universität des Saarlandes
D-66041 Saarbrücken, Germany

M. B. Panish, Ph. D.
AT&T Bell Laboratories
600 Mountain Avenue
Murray Hill, NJ 07974-2070, USA

Prof. R. M. Osgood
Microelectronics Science Laboratory
Department of Electrical Engineering
Columbia University
Seeley W. Mudd Building
New York, NY 10027, USA

Prof. H. Sakaki
Institute of Industrial Science
University of Tokyo
7-22-1 Roppongi, Minato-ku
Tokyo 106, Japan

Managing Editor: Dr. Helmut K. V. Lotsch
Springer-Verlag, Tiergartenstrasse 17
D-69121 Heidelberg, Germany

ISBN 3-540-57571-5 Springer-Verlag Berlin Heidelberg New York
ISBN 0-387-57571-5 Springer-Verlag New York Berlin Heidelberg

Library of Congress Cataloging-in-Publication Data. Laser ablation: principles and applications / John C. Miller (ed.); with contributions by L. L. Chase ... [et al.]. p. cm. — (Springer series in materials science; v. 28) Includes bibliographical references and index. ISBN 3-540-57571-5. — ISBN 0-387-57571-5 (U. S.) 1. Laser ablation. I. Miller, J. C. (John C.), 1949- . II. Chase, Lloyd L. III. Series. TA1715.L37 1994 621.36'6—dc20 94-5045

This work is subject to copyright. All rights are reserved, whether the whole or part of the material is concerned, specifically the rights of translation, reprinting, reuse of illustrations, recitation, broadcasting, reproduction on microfilm or in any other way, and storage in data banks. Duplication of this publication or parts thereof is permitted only under the provisions of the German Copyright Law of September 9, 1965, in its current version, and permission for use must always be obtained from Springer-Verlag. Violations are liable for prosecution under the German Copyright Law.

© Springer-Verlag Berlin Heidelberg 1994
Printed in Germany

The use of general descriptive names, registered names, trademarks, etc. in this publication does not imply, even in the absence of a specific statement, that such names are exempt from the relevant protective laws and regulations and therefore free for general use.

Typesetting: Macmillan India Ltd., Bangalore
SPIN: 10068717   54/3140-5 4 3 2 1 0 - Printed on acid-free paper

Preface.

Ever since the discovery of the laser in 1960, powerful beams of light have been directed at solid materials for a variety of purposes. Although not recorded, the first examples of laser ablation surely involved damage to optical components, either in the laser itself or elsewhere in the experimental arrangement. Later in the sixties, the process of laser ablation was studied and the roots of the major applications appeared. But it was not until the mid-eighties that the uses of laser ablation catapulted the technique into prominence. In particular, thin-film growth of high-T_c superconducting materials has been responsible for most of the papers in the field in the last eight to ten years. Other major applications such as laser medicine, laser-ablation mass spectrometry and other analytical techniques have also come to fruition in the late eighties. In the near future, additional novel uses of laser ablation such as X-ray generation, micromachining and lithography will accelerate.

Although there were focussed sessions or symposia on laser ablation at a number of conferences in the late eighties, the first meeting devoted solely to the topic was organized by researchers at Oak Ridge National Laboratory and Vanderbilt University, and was held in Oak Ridge in April 1991. The three-day workshop, sponsored by the U.S. Department of Energy, drew over eighty participants from five countries. The proceedings were published by Springer-Verlag in the Lecture Notes in Physics Series, Vol. 389, in the same year. The title, "Laser Ablation – Mechanisms and Applications", mirrored the emphasis on the fundamental physics and chemistry of laser ablation. Due to the success of this first workshop, a fully international conference was held two years later in April 1993.

The present book, like the workshop mentioned above, grew out of the realization that there was no single place that presented the recent progress in the diverse fields in which laser ablation was important. The invited speakers and participants of the Oak Ridge workshop represented a particularly prominent group of experts from which the present contributors were solicited. Following an introductory, historical chapter, the remaining chapters present an overview of a number of laser ablation applications with emphasis on the physical principles.

In Chap. 2, R.F. Haglund, Jr. (Vanderbilt University) and N. Itoh (Nagoya University, Japan) have outlined many of the theoretical ideas important to laser ablation with an emphasis on semiconductors and insulators. Although not coordinated with the other contributors, this chapter presents many ideas in detail which will be revisited briefly in later chapters.

L. Chase (Lawrence Livermore National Laboratory) discusses optical surface damage due to laser ablation in Chap. 3. Because of the importance of this area to the development of optics for high-power lasers, optical damage investigations have a long history.

In Chap. 4 the very important area of laser ablation of high-T_c superconductors is reviewed by T. Venkatesan (University of Maryland), a leading practitioner of this art.

R. Srinivasan (UV Technical Associates) has contributed a chapter on laser ablation of polymers. Ablation of these organic molecular solids is fundamentally different than that of inorganic materials.

Chapter 6 describes the coupling of laser ablation and mass spectroscopy, with an emphasis on biomolecules. R.L. Hettich and C. Jin (Oak Ridge National Laboratory) emphasize Fourier-transform mass spectrometry and describe work on inorganic clusters and on large molecules of biological interest.

The last (but of course not least) chapter, authored by A.D. Sappey and N.S. Nogar (Los Alamos National Laboratory), summarizes experiments emphasizing analytical applications of laser ablation.

In editing such a multidisciplinary collection of chapters authored by distinguished scientists from several countries there were a number of trials and tribulations, all of which were eventually overcome. I wish to thank all of the authors for their excellent efforts and for their tolerance of my nagging correspondence. Dr. H. Lotsch of Springer-Verlag gave valuable advice in planning and executing this project and was extremely patient as the deadlines were repeatedly broached. Finally, I wish to thank Ms D. Henderson and Ms N. Currence for their clerical support.

Oak Ridge, Tennessee *John C. Miller*
April 1994

Contents

1. History, Scope, and the Future of Laser Ablation
By *J.C. Miller* 1
1.1 Introduction 1
1.2 History of Laser Ablation Studies and Applications 2
    1.2.1 The Sixties 3
    1.2.2 The Seventies 6
    1.2.3 The Eighties 6
    1.2.4 The Nineties 8
References 9

2. Electronic Processes in Laser Ablation of Semiconductors and Insulators
By *R.F. Haglund, Jr.* and *N. Itoh* (With 18 Figures) 11
2.1 Electronic Mechanisms in Desorption and Ablation 12
2.2 Interaction of Photons with Solids 14
    2.2.1 Creation of Electron-Hole Pairs and Excitons 14
    2.2.2 Excitation of Electrons and Holes Localized on Defects 16
    2.2.3 Collective Effects: Free-Electron Heating and Plasma Effects 17
    2.2.4 Density of Electronic Excitation 17
2.3 Electron–Lattice Interactions and the Localized Excited State 18
    2.3.1 Interactions Between Free Carriers and Phonons ... 18
    2.3.2 Capture of Charge Carriers at Defect Sites 20
    2.3.3 Lattice-Induced Localization of Free Carriers and Excitons 21
2.4 Creation and De-Excitation of the Localized Excited State 23
    2.4.1 Non-Radiative De-Excitation 23
    2.4.2 Transfer of Electronic to Configurational Energy ... 25
    2.4.3 Other Non-Radiative De-Excitation Channels 26
2.5 Survey of Experimental Results 26
    2.5.1 Alkali Halides and Alkaline-Earth Fluorides 27

2.5.2 Oxides	28
2.5.3 Compound Semiconductors	30
2.6 Models of Laser-Induced Desorption	33
2.6.1 Models of Electronic Processes in Laser-Induced Desorption	33
2.6.2 Calculation Techniques	36
2.7 Simulation of Laser Ablation	39
2.7.1 Models of Laser Ablation	40
2.7.2 Model Calculations of Laser Ablation	42
2.8 Summary and Conclusions	47
References	48

3. Laser Ablation and Optical Surface Damage
By *L.L. Chase* (With 17 Figures)

	53
3.1 Introductory Remarks	53
3.2 Characteristics of Optical Surface Damage	55
3.3 Possible Causes of Optical Damage	58
3.4 Investigation of Optical Surface Damage Mechanisms	63
3.4.1 Laser Ablation as a Probe of Optical Damage	63
3.4.2 Surface Analytical Techniques	71
3.4.3 Laser Pump–Probe Measurements	74
3.5 Concluding Remarks	82
References	82

4. Pulsed-Laser Deposition of High-Temperature Superconducting Thin Films
By *T.V. Venkatesan* (With 17 Figures)

	85
4.1 Advantages of Pulsed-Laser Deposition	85
4.2 Materials Base	87
4.3 Laser-Beam–Target Interaction	88
4.3.1 Target Texturing	88
4.3.2 Particle Deposition	89
4.4 Dynamics of the Laser-Produced Plume	91
4.5 Evaporant–Substrate Interaction	93
4.6 Frontiers of High-Temperature Superconducting Thin-Film Research	94
4.6.1 Epitaxial Multilayers	94
4.6.2 Work on Ultrathin Films	96
4.6.3 Control of Phase and Crystallinity in Thin-Film Form	97
4.7 Scaling-up to Larger Areas	98
4.8 Future Directions	102
4.8.1 Component Development	102
4.8.2 System Issues	102
4.9 Summary	104
References	104

5. Interaction of Laser Radiation with Organic Polymers
By *R. Srinivasan* (With 21 Figures) 107
5.1 History. 107
5.2 Characteristics of UV-Laser Ablation 108
5.3 Chemical Physics of the Ablation Process 113
    5.3.1 Ablation Products 114
    5.3.2 Time Profile of Ablation 115
        a) Polyimide 116
        b) Polymethyl Methacrylate 123
5.4 Theories of Ultraviolet-Laser Ablation 129
5.5 Contemporary Trends in UV-Laser Ablation 130
References 131

6. Laser Ablation and Laser Desorption Techniques with Fourier-Transform Mass Spectrometry (FTMS)
By *R.L. Hettich* and *C. Jin* (With 7 Figures) 135
6.1 Principles of FTMS Operation. 136
    6.1.1 Ion Formation 136
    6.1.2 Ion Trapping. 137
    6.1.3 Ion Detection. 138
    6.1.4 Ion Structural Techniques 139
6.2 Laser-Ablation FTMS for Clusters 140
    6.2.1 Cluster Formation. 140
    6.2.2 Accurate Mass
          and High-Resolution Measurements 143
    6.2.3 Ion–Molecule Reactions 144
    6.2.4 Collision-Activated Dissociation. 146
6.3 Laser-Desorption FTMS for Biomolecules 147
    6.3.1 Development of Matrix-Assisted
          Laser Desorption 147
    6.3.2 Interfacing MALDI with FTMS 148
    6.3.3 Ion-Trapping Considerations for MALDI-FTMS . . . 149
    6.3.4 Combining Separation Methods
          with MALDI-FTMS 152
6.4 Future Directions. 152
6.5 Conclusions 154
References 154

7. Diagnostic Studies of Laser Ablation for Chemical Analysis
By *A.D. Sappey* and *N.S. Nogar* (With 5 Figures) 157
7.1 Laser Ablation in Vacuum 158
    7.1.1 Instrumentation 159
    7.1.2 Physical Processes for Laser Ablation In Vacuo 162
    7.1.3 Examples 165

7.2 Laser Ablation in an Atmosphere	166
7.2.1 Physical Processes Unique to Ablation in an Atmosphere	167
7.2.2 Diagnostics for Laser Ablation in an Atmosphere	171
a) Blast-Wave Diagnostics	172
b) Optical Diagnostics for Monitoring Plasma Formation	173
c) Density, Temperature, and Velocity Diagnostics	176
d) Ablated Material Velocity Determination	179
References	181
Subject Index	185

Contributors

Lloyd L. Chase

    Lawrence Livermore National Laboratory, P.O. Box 808, Livermore, CA 94550, USA

Richard F. Haglund, Jr.

    Department of Physics & Astronomy, Vanderbilt University, Nashville, TN 37235, USA

Robert L. Hettich

    Oak Ridge National Laboratory, Bldg. 4500S, Mail Stop 6120, P.O. Box 2008, Oak Ridge, TN 37831-6120, USA

Noriaki Itoh

    Department of Physics, Nagoya-U-Chikusaku, Nagoya 464, Japan

Changming Jin

    Oak Ridge National Laboratory, Bldg. 4500S, Mail Stop 6120, P.O. Box 2008, Oak Ridge, TN 37831-6120, USA

John C. Miller

    Oak Ridge National Laboratory, Bldg. 4500S, Mail Stop 6125, P.O. Box 2008, Oak Ridge, TN 37831-6125, USA

Nicholas Nogar

    Chemical & Laser Sciences Division, Los Alamos National Laboratory, CLS2 G738, Los Alamos, NM 87545, USA

A. D. Sappey

    Chemical & Laser Sciences Division, Los Alamos National Laboratory, CLS3 J565, Los Alamos, NM 87545, USA

Rangaswamy Srinivasan

UV Tech Associates, 2508 Dunning Drive, Yorktown Heights, NY 10598, USA

T. Venky Venkatesan

Superconductivity Center, University of Maryland, Physics Department, College Park, MD 20742, USA

1. History, Scope, and the Future of Laser Ablation

J.C. Miller

The study of the interaction of high-power lasers with solid matter is as old as the laser itself. Variously termed laser ablation, vaporization, sputtering, desorption, spallation or etching, the physical processes involved are extremely complex. Electromagnetic energy is converted to electronic, thermal, chemical and mechanical energy at the solid surface. The ejected material may include neutral atoms and molecules, positive and negative ions, clusters, electrons and photons. The generated plasmas may have electron temperatures of thousands of degrees. Of necessity a multidisciplinary problem, the mechanism of laser ablation is still being studied and debated.

Regardless of the detailed mechanisms, many important applications depend on laser ablation. These include industrial processes such as laser welding or hole drilling, materials processing to produce thin films or microstructures, elemental analysis of solid samples, biomedical uses such as laser surgery or structural studies of biomolecules and, finally, laser-based weapons such as those discussed in the Star Wars Initiative. A subset of these applications areas is represented in this volume and the relevant physical mechanisms are discussed in detail in the following chapters.

In the present chapter, the current status of laser ablation studies and applications will be put into a historical perspective and some future trends will be discussed.

1.1 Introduction

> "It is still a matter of wonder how the Martians are able to slay men so swiftly and so silently. Many think that in some way they are able to generate an intense heat in a chamber of practically absolute non-conductivity. This intense heat they project in a parallel beam against any object they choose, by means of a polished parabolic mirror of unknown composition, much as the parabolic mirror of a lighthouse projects a beam of light. But no one has absolutely proved these details. However, it is done, it is certain that a beam of heat is the essence of the matter. Heat, and invisible, instead of visible, light. Whatever is combustible flashes into flame at its touch, lead runs like water, it softens iron, cracks and melts glass, and when it falls upon water, incontinently that explodes into steam."
>
> H.G. Wells, *The War of the Worlds* (1896)

This prescient description of a directed energy weapon was published almost one hundred years ago [1.1]. It predated what we now know as the laser by 64 years. Yet the first paragraph above is immediately recognizable as a description of a laser. Also remarkable is the predicted effect such an intense beam of light would have on solid and liquid substances. The beam of "heat" would liquify, vaporize or combust the material it hits. That it would "crack and melt glass" is known to every present laser experimenter who uses a lens or a window in their experiment. The present volume is devoted in full to effects predicted in a book of science fiction!

Many terms have been used to refer to the processes occurring upon the interaction of high-power radiation with solid matter. Table 1.1 lists the most common. The terms used by researchers depend on their application area as well as the time period in which they worked. In the rest of this chapter, the term "laser ablation" will refer to the collective processes described by the various terms in Table 1.1.

Table 1.1. Terms commonly used for processes involving the interaction of high-power lasers with solid matter

Terms
Laser ablation
Laser vaporization
Laser desorption
Laser sputtering
Laser ejection
Laser etching
Laser spallation
Laser damage
Laser plasma generation
Laser-induced emission
Laser blow-off

1.2 History of Laser Ablation Studies and Applications

Although the requisite theoretical ideas were developed in the fifties with the emergence of the maser, the "laser age" is generally considered to have begun in 1960 with *Maimon's* publication demonstrating the first "optical maser" [1.2]. Among the first experiments using the new, intense light sources were those involving laser interactions with solids leading to the ejection of particles and photons.

A historical survey of these studies is conveniently divided into decades. Most of the current applications of laser ablation came to fruition in the eighties, but virtually all had their roots in research performed in the sixties. The literature became explosive after 1985, suggesting that the nineties will be an exciting period. A computer search of both Physics Abstracts and Chemistry Abstracts from 1970 to the present documents this trend. About 80% of the records found were dated after 1985.

1.2.1 The Sixties

As one might expect, the first archived accounts of laser ablation experiments appeared as abstracts of conference presentations. The very first such account known to the author is a presentation entitled "Optical Micromission [sic] Stimulated by a Ruby Maser" presented by *Breech* and *Cross* [1.3] at the International Conference on Spectroscopy held at the University of Maryland in June 1962. A focused ruby laser was used to vaporize and excite atoms from solid surfaces. The photon spectrum was dispersed and used to characterize the elements composing the surface. This paper began the field of laser microprobe emission spectroscopy which was the first real application area of laser ablation [1.4]. Commercial instruments appeared as early as the mid-sixties.

This early development was hastened by the existing technology for electron microprobe techniques [1.5]. Early application of the laser microprobe involved the elemental analysis of geological [1.6], biological [1.7] or metallic [1.8] samples. An interesting early use was the study of a forgery of a 15th century Flemish painting [1.9]. The analysis showed it to date from the 19th century. The first laser ablation of biological material involved the laser microprobe emission analysis of metallic elements in brain, pancreatic and other tissues [1.10]. Blood was similarly analyzed as early as 1964 [1.11]. Although the microprobe emission technique analyses photons, it depends, of course, on the ejection of material by the high-power laser.

The second (recorded) conference abstract in the field of laser ablation described experiments by *Linlor* [1.12] at Hughes Research Laboratories. In a talk entitled "Plasmas Produced by Laser Bursts", given at the American Physical Society (APS) Summer Meeting at the University of Washington in August 1962, *Linlor* ushered in an era of studying the physical mechanisms of laser ablation by describing measurements of the energy of ejected ions. Another very early conference talk by *Muray* [1.13] (Annual APS Meeting, January 1963) described the first study of photoelectrons emitted from glasses by a ruby laser pulse.

Throughout 1963 and 1964, about two dozen publications detailed the early experiments on laser ablation. Many phenomena which were first observed in those years are still the subject of study today. Furthermore, the techniques pioneered to investigate those phenomena have formed the basis for many widely used applications today. Details of these and other experiments performed during the sixties may be found in the book by *Ready* [1.14] entitled "Effects of High-Power Laser Radiation". Table 1.2 outlines a few of the important "firsts" which will be briefly described below.

Table 1.2. Important* "firsts" in laser ablation

Event	Researchers	References
First published account of laser ablation (measurement of photon emission)	Breech, Cross	[1.3]
First theoretical paper	Askar'yan, Moroz	[1.15]
First experimental paper (observation of electrons and positive ions; first use of mass spectrometry)	Honig, Woolston	[1.16]
First use of plume photography	Ready	[1.18]
First observation of clusters	Berkowitz, Chupka	[1.20]
First post-ionization of neutrals		
First ablation of biological material	Rosan et al.	[1.10]
First momentum-transfer experiment	Neuman	[1.25]
First suggestion of laser fusion	Basov, Krokhin	[1.26]
First VUV emission observation	Ehler, Weissler	[1.27]
First X-ray emission observation	Langer et al.	[1.28]
First observation of multiply-charged ions	Archibald	[1.29]
First two-photon photoemission	Sonneburg et al.	[1.30]
First three-photon photoemission	Logothetis, Hartman	[1.31]
First thin-film deposition	Smith, Turner	[1.32]

*In the opinion of the author

The first "regular" paper published was actually a theoretical contribution. *Askar'yan* and *Moroz* [1.15] calculated the recoil pressure during the ejection of laser ablated material and proposed the acceleration of small particles or droplets by means of "one-sided evaporation". They also suggested that ultrasonic and hypersonic oscillations resulting from modulated laser ablation should increase the material ejection.

The first experimental paper was a study by *Honig* and *Woolston* [1.16] of the laser ablation of several metals and semiconductors and of an insulator. They measured pulses of 3×10^{16} electrons and 10^8 positive ions from a single laser shot. About 2×10^{17} atoms were ablated, leaving a 150 μm diameter hole which was 125 μm deep. They analyzed the mass distribution with a modified commercial, double-focusing mass spectrometer, thus demonstrating the first use of ion microprobe analysis. The second published paper on laser ablation provided a more detailed study of the electron emission and its time profile. Assuming thermionic emission, *Lichtman* and *Ready* [1.17] estimated a surface temperature of 3 300 K for ruby laser interaction with tungsten. Possible multiphoton photoelectric effects were considered and discounted. In a later paper, *Ready* [1.18] employed high-speed photography to study the temporal and spatial profile of the ablated plume of ejected material. He showed that the plume from a carbon block emerged after the peak of the laser pulse and had its maximum brightness about 120 ns after the start of the laser pulse. The plume lasted several microseconds. The velocity of the plume's leading edge was estimated to be 2×10^6 cm/s. Again studying a carbon target, *Howe* [1.19] measured the vibrational (10000–20000 K) and rotational (4500 K) temperatures from the dispersed fluorescence of CN and C_2 molecules ejected from the sample. Interestingly, the different temperatures indicate the non-equilibrium conditions characteristic of cooling due to an adiabatic expansion. The first mention of such cooling and subsequent condensation was by *Berkowitz* and *Chupka* [1.20] who observed carbon ($n \leq 14$), boron ($n \leq 5$) and Mg ($n \leq 2$) cluster ions after post-ionization of the ablated plume. For the carbon clusters an even-odd alternation of intensities was observed as predicted by *Pitzer* and *Clementi* [1.21] and previously observed in a spark source [1.22]. Almost thirty years later, hundreds of papers on carbon clusters are being published annually and Buckyball (C_{60}) has been declared "The Molecule of the Year"[1.23]. It is interesting to speculate on the course of events had these early mass spectrometers been more sensitive at 720 amu [1.24]. The ablation of even larger "globules of molten material" and "splinters of material" was suggested by the first momentum transfer measurements [1.25].

By the end of 1964, virtually all of the essential first measurements on laser ablated solids had been made by monitoring various properties of electrons, ions, neutrals, clusters and photons. Estimates of temperatures, velocities and kinetic energies had been made. A coherent picture of the process was emerging. The rest of the sixties saw the extension and systematization of these early observations to include numerous other targets. Studies were performed over a wide range of laser characteristics such as power, pulse width, etc. The lasers themselves and the measurement techniques became increasingly sophisticated which allowed more detailed theoretical descriptions. In particular, ever-higher laser powers were employed and hotter plasmas studied. The first suggestion of laser fusion by *Basov* came in 1964 [1.26]. Later in the decade VUV [1.27] and X-ray emissions [1.28] were measured, supplementing the earlier visible and near UV observations. Multiply charged ions [1.29] were observed and two- [1.30] and three-photon [1.31] photoemissions were discovered as laser power increased. Further details of these early experiments may be found in various reviews [1.4–8, 14].

Of particular importance, in view of the explosive study of high-temperature superconductors in the late eighties, the first experiment demonstrating the laser deposition of thin films was performed by *Smith* and *Turner* [1.32] in 1965. These authors ablated a variety of materials with a ruby laser and demonstrated that thin films could be grown in this way. This application area was slow to develop as other sputtering techniques produced better quality films. Not until the eighties was laser film growth able to compete with films grown by other techniques, such as Molecular Beam Epitaxy (MBE).

Along with the experimental and theoretical understanding of laser ablation, the first real applications of the technique were developed in the sixties. By the end of the decade, laser microprobe emission and mass spectroscopy were in routine use for elemental analysis [1.5]. Early industrial applications of high-power lasers such as to cutting, welding, scribing and hole drilling were described by *Ready* [1.14].

1.2.2 The Seventies

This decade represented a time of expanded uses of laser ablation for analysis of various materials and also more detailed studies of the physics of the ablation. Much of this research was driven by improvements in laser technology. Microanalysis of solid samples remained the major application of laser ablation. In particular, the use of mass spectrometric analysis grew rapidly in the latter half of the seventies, culminating in the first commercial instrument introduced by Leybold-Heraeus in 1978. Progress in this area is detailed in several chapters of the book "Lasers and Mass Spectrometry" edited by *Lubman* [1.33]. However, laser microprobe emission spectroscopy remained the more popular technique. Post-excitation and post-ionization techniques were explored to analyze the neutrals ejected from the surface. Greater sensitivity and selectivity were achieved in this way. The review by *Cremers* and *Radziemski* [1.34] is a particularly useful source for work performed during the seventies and early eighties.

Again, thin-film development was slow in the seventies, hampered by the poor quality of the films. The growing reliability and stability of commercial lasers, particularly *Q*-switched YAG lasers, improved the uniformity of film growth as well as improving the reproducibility of microprobe measurements.

The introduction of dye lasers in the mid-seventies had little effect on the field of laser ablation, as resonance effects, which are usually broad in energy, were seldom studied. Plasmon effects and resonant post-ionization techniques were not explored until the eighties.

Most progress in the seventies on fundamentals concentrated on plume diagnostics and model development. More detailed studies of the mass and charge dependence of the ejected plume were carried out under varied laser conditions and on a vast array of solid and liquid surfaces. The generated plasma was characterized by its temperature as measured by the kinetic energy of the electrons, ions and neutrals. By the end of the seventies a broad understanding of the physics of laser ablation was achieved.

1.2.3 The Eighties

The growth of the literature on laser ablation in the eighties (especially the latter half) has been explosive. Computer searches under key words such as laser ablation or laser desorption typically yield 5–10 times more citations since 1985 than the total of all papers published in previous years. This growth is based on two factors – the expansion and integration of laser technology into virtually every scientific laboratory and the increased number and importance of the various applications. The technique of laser ablation, like the laser itself, could be characterized as an "answer in search of a question". The "questions" multiplied rapidly in the eighties, and the scientific community responded in force.

Lasers became ubiquitous in the eighties, and all of their important characteristics were pushed to new limits. Picosecond laser systems became common and femtosecond lasers emerged. High peak-power lasers could be focused to produce 10^{16} W/cm^2 on a target (ultra-large lasers such as NOVA could exceed this). Tunable lasers expanded their range into the ultraviolet and infrared. The Nd:YAG and CO_2 lasers remained the mainstay for ablation studies, but the excimer laser emerged as a major player in the eighties. Powerful, pulsed tunable dye lasers were readily available to most experimenters. Multiple-laser experiments became common. Concurrently, the sophistication of other laboratory equipment, particularly electronics and computers, vastly increased the experimenters' ability to acquire and analyze data. Mass and electron spectrometers, molecular beam machines, ultra-high vacuum equipment, high-speed cameras, fiber optics, etc. became cheaper and more capable.

Along with the dizzying array of new technology, applications of laser ablation became more mainstream and economically useful, thus drawing more scientists into the field.

Undoubtedly, the fastest growing applications in the decade of the eighties were driven by the needs of the materials sciences. The development of thin films grown by pulsed-laser deposition reached the stage where unique films could be created by the laser technique. The advent of high-T_c superconductors drove hundreds of laboratories to investigate thin films. The observation that laser ablation transfers the stoichiometry of a multicomponent system to the film and allows oxygenation and other chemistry to occur in transit from target to substrate is truly remarkable. Virtually all materials – metals, semiconductors, and insulators – can be deposited. Laser-assisted chemical vapor deposition combines gas-phase chemistry with surface physics.

Another driving force for laser ablation was in the area of micron and submicron structures. Ever faster computers require smaller and smaller elements. Lithography and micromachining reached a new size regime and have spawned the field of nanotechnology. References and reviews of laser ablation and materials applications in the eighties may be found in proceedings of the annual meetings of the Materials Research Society. A useful review was authored by *Cheung* and *Sankur* [1.35].

Another new application of laser ablation came in this decade in the field of medicine. Laser surgery progressed from research to routine in this period. Laser ablation of tissue, of arterial blockages, and of tumors and dental material were all successfully demonstrated. Laser surgery of the eye has been perhaps the most common among these techniques. Understanding of the physics and chemistry, and now biology, of laser–material interaction has been crucial to the introduction of such new medical procedures.

Bridging laser surgery and laser analytical analysis has been the growing use of laser mass spectroscopy for the study of biomolecules. Although, it is not possible to identify a "first", early papers by *Hillenkamp* et al. [1.36] and *Posthumus* et al. [1.37] clearly set the stage for the studies of the eighties. Several major issues were addressed during the eighties. Simply obtaining the necessary gas-phase density of low vapor-pressure biomolecules without fragmentation was the first challenge. Various matrix- or substrate-assisted ablation techniques were developed to solve this problem. To prevent fragmentation during ionization, near threshold techniques were used to achieve what was called "soft ionization". Characterization of large biomolecules also required increases in resolution and mass range. The use of Fourier Transform Mass Spectrometry (FTMS) and Time-of-Flight (TOF) techniques (especially reflectrons) solved these problems. Major funding initiatives by the Department of Energy and the National Institutes of Health involving structural biology and genome sequencing led to rapid growth in laser desorption mass spectrometry of these large molecules.

Another major new application of ablation techniques arose in the eighties. This was the study of clusters formed during isentropic expansions. Although laser ablation into vacuum was shown very early to lead to clustering [1.20], the development of pulsed, laser ablation supersonic nozzle sources allowed a wide range of non-volatile materials to be cooled in a rare gas expansion. The resultant clusters were then examined, usually by mass spectrometry. The first use of this type of cluster source is generally credited to *Smalley* and co-workers [1.38]. The interest in clusters grew steadily through the eighties and laser ablation sources were constructed by dozens of researchers. Initially only one of many interesting studies in the eighties [1.39, 40], the observation of special stability of carbon clusters containing sixty or seventy atoms resulted in a "feeding frenzy" by the end of the decade. The study of Buckyball (C_{60}) or, in general, fullerenes and their related substituted compounds has dominated both the technical and popular literature. For a review of the field a special issue of Accounts of Chemical Research is recommended [1.41].

The growing use of lasers in the defense industry fueled several areas related to laser ablation. In particular, the Strategic Defense Initiative (SDI), established in 1982, led to much research on directed energy weapons, including high-power lasers. This led to a new interest in optical surface damage and in the effect of such intense laser beams on missiles or satellites in space. Of course, the field of optical surface damage must surely date back to the first lens or window placed in front of an early (perhaps the first) ruby laser. But the growth of laser-based weapons clearly encouraged much research in laser damage. This work may be followed in the literature through the proceedings of the "Boulder Damage Conference" published by the National Bureaus of Standards (now the National Institute of Science and Technology) as a special publication series entitled "Laser-induced Damage in Optical Materials" (see Chap. 3 and references therein).

Another ablation application which may blossom in the nineties was also accelerated by SDI. The development of X-ray lasers generally begins with the generation of a dense plasma by laser ablation, leading to amplified emission in the VUV and X-ray region by multiply-charged ions. The successful development of a tabletop X-ray laser would certainly have a large impact on fields like lithography and biological imaging.

1.2.4 The Nineties

We begin the nineties with the certainty that laser ablation and its applications will play a role in many areas of scientific research and development. A glance at the latest conference proceedings in any of these fields will demonstrate the ubiquitous nature of the technique. In 1991, the first conference devoted solely to laser ablation was hosted by the Oak Ridge National Laboratory [1.42] and a similar European meeting was held later in the same year [1.43]. The present volume arose out of the success of the Oak Ridge conference by providing a stable of prominent potential chapter authors. The second International Conference on Laser Ablation (COLA 93) was held in the Spring of 1993 [1.44].

All of the applications described above will certainly flourish and mature in the nineties. It remains to be seen what new applications may emerge. Or perhaps a new advance in laser technology may fundamentally change the present applications. However the outcome, the nineties will be an exciting decade.

References

1.1 H.G. Wells: The War of the Worlds (Bantam Doubleday Dell, New York 1988)
1.2 T.H. Maimon: Nature **187**, 493 (1960)
1.3 F. Breech, L. Cross: Appl. Spectrosc. **16**, 59 (1962)
1.4 I. Harding-Barlow, K.G. Snetsinger, K. Keil: Laser Microprobe Instrumentation, in *Microprobe Analysis*, ed. by C.A. Andersen (Wiley, New York 1973)
1.5 C.A. Andersen: *Microprobe Analysis* (Wiley, New York 1973)
1.6 K.Keil, K.G. Snetsinger: Applications of the Laser Microprobe to Geology, in *Microprobe Analysis*, ed. by C.A. Andersen (Wiley, New York 1973)
1.7 I. Harding-Barlow, R.C. Rosan: Application of the Laser Microprobe to the Analysis of Biological Materials, in *Microprobe Analysis*, ed. by C.A. Andersen (Wiley, New York 1973)
1.8 M. Margoshes: Application of the Laser Microprobe to the Analysis of Metals, in *Microprobe Analysis*, ed. by C.A. Andersen (Wiley, New York 1973)
1.9 Anonymous: Laser Focus **1**, 8 (1965)
1.10 R.C. Rosan, M.K. Healy, W.F. McNary, Jr.: Science **142**, 236 (1963)
1.11 N.A. Peppers: Appl. Opt. **4**, 555 (1965)
1.12 W.I. Linlor: Bull. Am. Phys. Soc. **7**, 440 (1962)
1.13 J.J. Muray: Bull. Am. Phys. Soc. **8**, 77 (1963)
1.14 J.F. Ready: *Effects of High-Power Laser Radiation* (Academic, New York 1971)
1.15 G.A. Askar'yan, E.M. Moroz: Zh. Eksp. Teor. Fiz. **43**, 2319 (1962) [English transl. Soviet Phys.-JETP **16**, 1638 (1963)]
1.16 R.E. Honig, J.R. Woolston: Appl. Phys. Lett. **2**, 138 (1963)
1.17 D. Lichtman, J.F. Ready: Phys. Rev. Lett. **10**, 342 (1963)
1.18 J.F. Ready: Appl. Phys. Lett. **3**, 11 (1963)
1.19 J.A. Howe: J. Chem. Phys. **39**, 1362 (1963)
1.20 J. Berkowitz, W.A. Chupka: J. Chem. Phys. **40**, 2735 (1964)
1.21 K.S. Pitzer, E. Clementi: J. Am. Chem. Soc. **81**, 4477 (1959)
1.22 E. Dörnenburg, H. Hintenburger: Z. Naturforsch. **14a**, 765 (1959)
     E. Dörnenburg, H. Hinterburger, J. Franzen: Z. Naturforsch. **16a**, 532 (1961)
1.23 D.E. Koshland, Jr.: Science **254**, 1705 (1991)
1.24 W.A. Chupka: Private communication (1992)
1.25 F. Neuman: Appl. Phys. Lett. **4**, 167 (1964)
1.26 N.G. Basov, O.N. Krokhin: Zh. Eksp. Teor. Fiz. **46**, 171 (1964) [English transl.: Soviet Phys. - JETP **19**, 123 (1964)]

1.27 A.W. Ehler, G.L. Weissler: Appl. Phys. Lett. **8**, 89 (1966)
1.28 P. Langer, G. Tonen, F. Floux, A. Ducauze: IEEE QE-2, 499 (1966)
1.29 E. Archibald, T.P. Hughes: Nature **204**, 670 (1964)
1.30 H. Sonneburg, H. Heffner, W. Spicer: Appl. Phys. Lett. **5**, 95 (1964)
1.31 E.M. Logothetis, P.L. Hartman: Phys. Rev. Lett. **18**, 581 (1967)
1.32 H.M. Smith, A.F. Turner: Appl. Opt. **4**, 147 (1965)
1.33 D.M. Lubman (ed.): *Lasers and Mass Spectrometry* (Oxford Univ. Press, Oxford 1990)
1.34 D.A. Cremers, L.J. Radziemski: Laser Plasmas for Chemical Analysis, in *Laser Spectroscopy and Its Applications*, ed, by L.J. Radziemski, R.W. Solarz, J.A. Paisner (Marcel Decker, New York 1987)
1.35 J.T. Cheung, H. Sankur: CRC Crit. Rev. Solid State Mater. Sci. **15**, 63 (1988)
1.36 F. Hillenkamp, E. Unsöld, R. Kaufmann, R. Nitsche: Nature **256**, 119 (1975)
1.37 M.A. Posthumus, P.G. Kistemaker, H.L.C. Meuzelaar, M.C. Ten Noever de Brauw: Anal. Chem. **50**, 985 (1978)
1.38 T.G. Dietz, M.A. Duncan, D.E. Powers, R.E. Smalley: J. Chem. Phys. **74**, 6511 (1981)
1.39 E.A. Rohlfing, D.M. Cox, A. Kaldor: J. Chem. Phys. **81**, 3322 (1984)
1.40 H.W. Kroto, J.R. Heath, S.C. O'Brien, R.F. Curl, R.E. Smalley: Nature **318**, 162 (1985)
1.41 Accounts of Chemical Research **25** (1992)
1.42 J.C. Miller, R.F. Haglund, Jr. (eds.): *Laser Ablation Mechanisms and Applications*, Lect. Notes Phys., Vol. 389 (Springer, Berlin, Heidelberg 1991)
1.43 E. Fogarassy, S. Lazare (eds.): *Laser Ablation of Electronic Materials*, European Materials Research Society Monographs, Vol. 4 (North-Holland, Amsterdam 1992)
1.44 J.C. Miller, D.B. Geohegan (eds.): *Laser Ablation: Mechanisms and applications-II*, AIP Conference Proceedings **288** (American Institute of Physics, New York 1994)

2. Electronic Processes in Laser Ablation of Semiconductors and Insulators

R.F. Haglund, Jr. and N. Itoh

With 18 Figures

Surface ablation by high-power, nanosecond-pulsed lasers is playing an increasingly important role in materials processing technologies [2.1] and it is a matter of significant scientific interest in its own right. Laser ablation in the nanosecond regime has traditionally been viewed as resulting from rapid heating of the surface layer on the grounds that typical thermalization times, even in insulators, are much shorter than the laser-pulse duration. While the interaction of laser light with metal surfaces in ultra-high vacuum has thermal characteristics [2.2], it is increasingly clear that electronic effects play a significant, and possibly dominant, role in laser ablation from nonmetallic solid surfaces. For example, color centers or other electronic defects may be created by the initial laser irradiation, changing the absorption characteristics – and thus the deposition of laser energy – for photons late in a given laser pulse or in subsequent laser pulses. Diffusion of electronic defects to the surface weakens atomic bonds, resulting in preferential ejection or evaporation of certain atomic or ionic species. Surface-conditioning or incubation effects occur even for interpulse spacings which are clearly much longer than characteristic thermal diffusion times. The challenge, in the light of the varied phenomenology of laser ablation in nonmetallic materials, is to identify the common underlying mechanisms.

In this review, we focus on the electronic processes which underly nanosecond pulsed-laser ablation in pure semiconductor and insulating materials. We do not treat laser-induced desorption and photochemistry of surfaces with adsorbed molecules [2.3]; we shall also leave aside the intricate interplay of photophysics and chemistry in matrix-assisted laser-induced desorption and ionization, now a major analytical technique in studies of proteins and other macromolecules [2.4]. Instead, we attempt to develop a framework for modeling both the kinetics and dynamics of laser-induced desorption and ablation from nonmetallic crystalline surfaces. We first characterize laser–solid interactions, then analyze the processes which lead to defect formation, particle emission and, finally, ablation. After a survey of laser-induced desorption and ablation experiments on halide, oxide and semiconductor surfaces, we then consider models and computations of both desorption and ablation in the limiting cases of solids with strong and weak electron–lattice coupling.

Specialized features of this subject have been treated in recent reviews, including laser ablation of insulators [2.5] comparisons of laser ablation and heavy particle sputtering [2.6] and modification of solids by high-intensity lasers [2.7]. Bulk and surface electronic processes in insulating and semi-conducting solids have been reviewed by *Itoh* [2.8].

2.1 Electronic Mechanisms in Desorption and Ablation

As a working definition, we take laser ablation to exhibit *all* of the following characteristics: (*i*) material removal rates exceeding one monolayer per pulse with residual nonstoichiometric surface; (*ii*) atom, ion or cluster yields which are superlinear functions of the electronic excitation density; and (*iii*) a fluence threshold below which only particle emission without destruction of the surface is observed [2.9]. While we shall not consider it explicitly in this review, we note that laser ablation at high fluences is accompanied by formation of a weakly ionized plasma which gives rise to interesting phenomena through plasma–surface interactions and laser-induced plasma photochemistry. Where particle emission occurs without damage of the surface, we shall refer instead to laser-induced desorption or laser-induced particle emission. Laser-induced particle emission must be considered in any discussion of laser ablation, since it frequently generates the defects which are precursors to laser ablation.

The electronic mechanisms operative in laser-induced desorption and ablation are shown in Fig. 2.1. The primary interaction of photons with surfaces produces electron-hole pairs, heats free electrons and may generate local heating

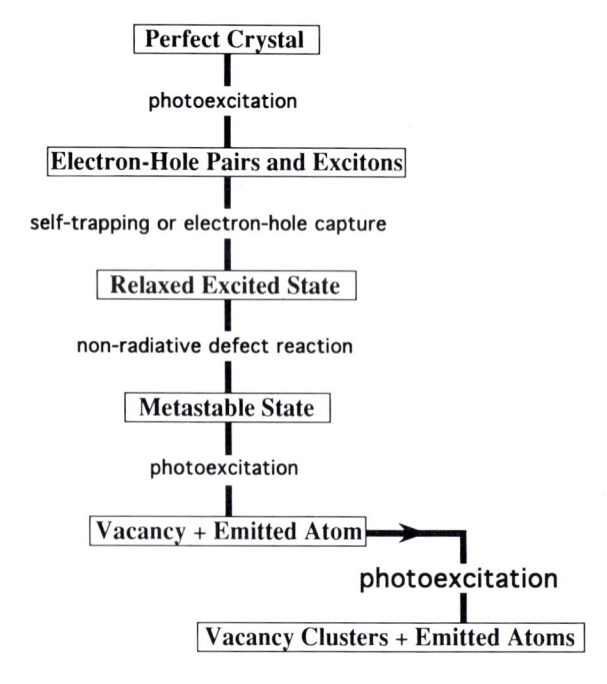

```
        Perfect Crystal
              |
        photoexcitation
              |
   Electron-Hole Pairs and Excitons
              |
  self-trapping or electron-hole capture
              |
       Relaxed Excited State
              |
    non-radiative defect reaction
              |
         Metastable State
              |
         photoexcitation
              |
     Vacancy + Emitted Atom ──┐
                              |
              photoexcitation
                              |
     Vacancy Clusters + Emitted Atoms
```

Fig. 2.1. Schematic diagram of electronic processes leading to laser ablation

around optically absorbing centers. Two key physical parameters which characterize the primary mechanism are the *density of electronic excitation* and the *flux density* of the photon field. The former is the volumetric density of electron-hole pairs created by the incident laser light while the latter is the volumetric density of photons per unit time capable of exciting products of the primary photon–solid interaction.

Laser-induced desorption and ablation result when the initial electronic excitation is converted into the driving energy for nuclear motion, resulting in the ejection of atoms, ions and molecules from the surface. The rate and dynamical features of this conversion depends critically on *electron–lattice interactions* characteristic of the laser-irradiated solid, such as scattering of free electrons by phonons, phonon emission, localized lattice rearrangements and configuration changes. This last category includes: self-trapping of holes and excitons, defect formation and defect reactions, and surface decomposition due to electronic interactions of defects with lattice ions.

Accompanying or following the electron–lattice interaction are secondary electronic processes such as: photoabsorption by free electrons, successive excitations of self-trapped excitons and electronic defects, transient changes in optical absorption as the surface layer or near-surface bulk is decomposed, photoemission and ionization of neutral species by the incident laser light, generation of a laser-ablation plasma and photoacoustic or photothermal processes. We propose that it is principally these secondary processes which generate the characteristic superlinear dependence of yield on electronic excitation density in laser ablation.

In summary, our conceptual framework for a physical model of laser ablation based on electronic processes incorporates:

— the primary interaction mechanism of photons with solid surfaces,
— electronic, vibrational and configurational electron–lattice relaxation,
— the evolution of defect and other characteristic vacancy clusters, and
— secondary energy-dissipation processes which produce the catastrophic nonlinear response typical of laser ablation.

On a nanosecond time scale, both the optical properties of the laser-irradiated material and the interaction and relaxation mechanisms are expected to change during and after the laser pulse. The combination of changed optical absorption characteristics and the evolution of defect clusters in the surface produces large numbers of weakly bonded atoms which can be dislodged at relatively modest laser fluences.

The initial photon–surface interaction produces defects both by altering the electronic and geometrical structure of the surface and bulk, and by laser-induced particle emission. As these defects grow in number, during one or multiple successive laser pulses, defects and defect clusters evolve on the surface and in the sub-surface bulk. The changing optical, topological and structural properties of the defect clusters then create the conditions for massive ejection of surface atoms, molecules and clusters. This exceedingly general mechanism operates in insulators and semiconductors alike; the difference from one material to another lies solely in the route taken to produce the defect clusters in the first place.

2.2 Interaction of Photons with Solids

The primary interaction of photons with solids induces electronic transitions from the bulk valence band, from occupied surface and defect states, and from free conduction-band states. The precise character of this initial electronic excitation depends on both the photon energy and the local electronic structure of the solid. The optical absorption coefficient of solids for photons with energies greater than the bulk bandgap typically exceeds 10^5 cm^{-1} if the transition near the band edge is direct. Thus, heating is often the primary concern for laser irradiation by photon energies of this range, and this has led to the common use of the rubric "thermal ablation" to describe these processes. Typical cross sections of each excitation mode for photons below and above the band-to-band transition energies are given in Table 2.1.

2.2.1 Creation of Electron-Hole Pairs and Excitons

The optically excited states of a non-metallic solid include electron-hole pairs, exitons or both. The exiton may be considered as an electron–hole pair with an energy lower than the conduction band edge or, more rigorously, as an excited state of the entire electronic system in a solid [2.10]. Photons with an energy hν above the energy E_{exc} of the first exciton peak produce both electron–hole pairs

Table 2.1 Photon–Solid Interaction Mechanics and Cross Sections

Photon energy	Initial electronic transition	Approximate cross section $\sigma^{(n)}$ [cm^{2n} s$^{-(n-1)}$]
$h\nu > E_{gap}$	Single photon band-to-band transitions in bulk solid	~ 10^{-17} for direct transitions, smaller for indirect transitions
	Transitions involving surface states	~ 10^{-17} for direct transitions, smaller for indirect transitions
$h\nu < E_{gap}$	Multiphoton band-to-band transitions of order n (resonant or nonresonant)	$\sigma^{(2)} \sim 10^{-34} - 10^{-50}$ $\sigma^{(3)} \sim 10^{-51} - 10^{-80}$ $\sigma^{(4)} \sim 10^{-68} - 10^{-140}$ Resonant transitions are larger than non-resonant by factors of $10^{4n} - 10^{8n}$
	Single-photon transition involving the surface state	~ 10^{-17} for direct transition, smaller for indirect transition
	Single-photon transition involving the defect state	~ $10^{-20} - 10^{-16}$

and excitons: those between the first exciton peak and the band-gap energy E_{gap} produce excitons, while those above E_{gap} produce electron–hole pairs. Surplus kinetic energy, $(hv-E_{exc})$ for excitons and $(hv-E_{gap})$ for electron–hole pairs, is transferred to phonons by electron–lattice interactions within a picosecond. Thus, absorption of a photon produces both electronic and vibrational excitation in the surface and near-surface regions of the irradiated solid.

In insulators, which have relatively small dielectric constants, excitons are strongly bound. In semiconductors, on the other hand, excitons are converted immediately to electron–hole pairs, except at very low temperatures. Therefore, dense electronic excitation of a semiconductor produces an electron–hole plasma, while dense electronic excitation of an insulator yields a mixture of electron–hole pairs and excitons. Electron–hole pairs are converted to excitons on a time scale of the order of $\tau_c \propto Ke/(8\pi\mu n)$, where K is the dielectric constant, e the electronic charge, μ the mobility and n the concentration [2.11]. This time constant is often smaller than 1 ps. This implies that, in typical insulators, excitons will predominate over free electron–hole pairs.

In an electron–hole plasma, each charge carrier is surrounded by carriers of opposite charge, reducing the Coulomb repulsion between carriers of like sign by a screening factor $\exp(-kr)$, where k, the screening constant, varies as $n^{1/6}$, and r is the distance between the carriers [2.12]. The electron–hole plasma may exhibit ambipolar diffusion [2.13] in which the local density of the electrons and holes are strongly correlated.

The electron–hole plasma in semiconductors can be de-excited by Auger recombination and by radiative and non-radiative recombination at defects. The cross section of the former is inversely proportional to the density n^3 of electron–hole pairs, and hence it dominates at extremely high concentration. Recombination at a defect occurs through a capture of an electron followed by another capture of a hole, and vice versa [2.14]. The electron is first captured into an excited state close to the continuum; the excitation energy is subsequently dissipated by phonon emission. The second particle is de-excited non-radiatively to the lowest excited state and to the ground state by photon or phonon emission.

Excitons in insulators de-excite through both radiative and non-radiative recombination channels. In solids in which excitons couple weakly with the lattice, a sharp free-exciton luminescence band is observed. In solids with strong electron–lattice coupling, luminescence with a large bandwidth from self-trapped excitons is observed [2.15, 16]. We shall see later that the characteristics of laser ablation depend strongly on whether self-trapping occurs or not.

Multiphoton band-to-band transitions also produce electron–hole pairs and excitons, just as do single-photon transitions. However, since the volume density of electronic excitation generated by multiphoton transitions is extremely low, the number of electron–hole pairs and the desorption yields resemble those produced by one-photon band-to-band transitions by much weaker light sources, such as electron storage rings. It has been suggested that multiphoton band-to-band transitions can be resonantly enhanced by defect states, greatly

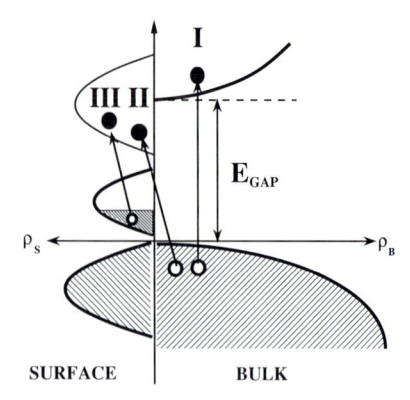

Fig. 2.2. Schematic of accessible electronic transitions at solid surfaces: I is a band-to-band transition; II is a transition from the valence band to a surface state; and III is a transition from the surface state to the conduction band

increasing the density of electronic excitation and the number of electron–hole pairs [2.17]. The probability for this resonant enhancement depends on the electronic structure of the solid, the precise character of the defects involved in the excitation and the laser wavelength. Once produced by multiphoton band-to-band transitions near the surface, defects may subsequently be excited by single- or multi-photon transitions.

The initial excitation from bulk valence-band states to the conduction band produces three-dimensional electron–hole pairs, while excitations involving surface states produce two-dimensional electron–hole pairs. The latter may involve transitions between occupied and unoccupied surface states, between the valence band and unoccupied surface states, or between the occupied surface states and the conduction band, as shown in Fig. 2.2. The connection between the two-dimensional electron–hole pairs and laser-induced desorption and ablation has been demonstrated very recently [2.18, 19]. The electron–hole pairs produced by these transitions are initially constrained to the two-dimensional surface but may be converted to three-dimensional electron–hole pairs by scattering with phonons.

If the energy gap for the transitions involving surface states $E_{\text{gap}}^{\text{surf}}$ is smaller than E_{gap} for the bulk, surface-state transitions can be excited by photons whose energy is too small to excite bulk band-to-band transition as, for example, in GaP where there is an unoccupied surface state 0.7 eV below the bulk band edge [2.20]. Two-dimensional electron–hole pairs created with photons having $h\nu \sim E_{\text{gap}}^{\text{surf}}$ are confined in two-dimensional surface orbitals for an extended time, since conversion to the three-dimensional electron–hole pairs is energetically forbidden.

2.2.2 Excitation of Electrons and Holes Localized on Defects

Electronic defects in the bulk of solids generate discrete levels in the band gap, as do the defects on the surface. Optical transitions involving these defect levels will dominate for photons of sub-gap energies, as noted in the voluminous work on optical excitation of defects in the bulk [2.21]. Optical transitions between localized defect orbitals are often observed. The cross section for these transitions depends on the transition matrix element between the defect ground state and excited state; for allowed transitions it is of the order of 10^{-17} cm^2. The transition from a defect-localized orbital to a continuum is also observed; in this case, optical absorption cross sections are smaller. Not much information is available on the optical transitions of defects on surfaces. It has recently been shown that excitation of defects on surfaces is associated with desorption and ablation in both Al$_2$O$_3$ [2.22] and GaP [2.23]. When an electron is trapped at an excited state of a defect which has previously trapped a hole (or vice versa), the same excited state as that produced by electronic excitation of a defect may be produced in the electron–hole recombination process.

2.2.3 Collective Effects: Free-Electron Heating and Plasma Effects

Optical absorption by free electrons in an electron–hole plasma can lead to heating through electron–phonon scattering [2.24]. In NaCl, a free-electron concentration of 10^{18} in NaCl can be generated by multiphoton band-to-band transitions at a flux of 5×10^{29} cm^2 s^{-1} [2.25]. Rapidly heated by single-photon absorption, these electrons heat the crystal by electron–phonon scattering to a few hundred degrees, producing local stresses exceeding the yield strength. A recent theoretical study of free-electron heating in SiO$_2$, using Monte-Carlo techniques to integrate the Boltzmann transport equations in an intense radiation field, confirms the role of multiphoton absorption by free carriers in heating the lattice in this wide bandgap insulator as well [2.26]. This study finds that the laser–electron interaction must be treated quantum mechanically at visible wavelengths, while a classical treatment in terms of dielectric breakdown is appropriate for wavelengths longer than 2 μm. More importantly, this approach makes it possible to incorporate realistic scattering rates and to predict experimentally observable indicators of laser damage, at least in the bulk.

2.2.4 Density of Electronic Excitation

The density of electronic excitation plays a major role in determining the response of a solid to photon irradiation. The number of excitations per second of a specific site in a solid is given by $N^{(n)} = \sigma^{(n)}(I/h\nu)^n$ for multiphoton excitation of order n, where σ is the excitation cross section and I is the irradiance, usually stated in W/cm^2. Using the cross sections listed in Table 2.1, the number of band-to-band excitations for a single site in a nanosecond pulse with a fluence of 1 J cm^2 is of the order of 10^2 for single-photon excitation, but of the order of 10^{-7}–10^{-10} for two- and three-photon excitations.

Single-photon band-to-band transitions in the bulk produce electron–hole pairs with a spatial density $\varrho = \sigma I n \tau/(h\nu)$, where τ is the lifetime of electron–hole pairs and n is the number of atoms per unit volume. The energy of the electron–hole pairs, ultimately dissipated to phonons by recombination, heats the lattice at a rate of $\sigma \Phi n$ per unit time and per unit volume, where

$\Phi = I/h\nu$ is the laser flux density in units of cm^{-2}s^{-1}. When electronic transitions are limited to surface states, heat is generated only at the surface and transported into the bulk; thus, heating effects in this case may be insignificant. When defects are excited, electronic excitation is transferred to local lattice modes and the temperature rise is, once again, not a dominant effect. When the photon flux is high, cascade excitation of surface defects can cause emission of atoms from defect sites [2.27].

2.3 Electron–Lattice Interactions and the Localized Excited State

In non-metallic solids, the electron–hole pair created by photon absorption relaxes to form a localized and excited electronic and vibrational state. The conversion of the electronic energy of the absorbed photon to nuclear motion and ultimately to desorption is determined by the character of electron–lattice interactions. Of course, not all electronic excitations lead to desorption; an electron–hole pair may lose energy or recombine radiatively and be lost from the precursor channels for desorption. The various excitation and relaxation pathways for the transition from the primary photon–solid interaction to the relaxed excited state are shown schematically in Fig. 2.3.

In general, an electron or a hole created by photoexcitation of a solid can relax or reduce its energy through both delocalized and localized electron–lattice interactions. Delocalized electron–phonon interactions are chiefly of concern in transport theory; they include [2.13]: the Fröhlich or polaron interaction with transverse optical phonon modes, the piezoelectric interaction with the acoustic modes and the deformation potential. Localized electron–lattice interactions, on the other hand, have spatially compact wave functions of the type which can lead to desorption and ablation. In this section, we consider scattering, capture and lattice-induced localization.

2.3.1 Interactions Between Free Carriers and Phonons

The polaron is an elementary excitation produced by absorption of a photon and consists of the photo-generated carrier dressed by its cloud of accompanying phonons. Polaron size is remarkably dependent on electron–lattice coupling strength. One usually speaks of "large" polarons with radii of the order of 10–15 Å, and of "small" polarons with radii a factor of two or three smaller. In weak-coupling materials such as compound semiconductors, scattering is more probable than small-polaron formation or trapping, whereas the reverse is true for strong-coupling solids. Scattering of free carriers is important to desorption and ablation in two ways. First, inelastic scattering slows carriers to the point where they may be captured, thus creating electronic defects which may lead to desorption/ablation. Second, scattering of laser-excited free carriers by phonons

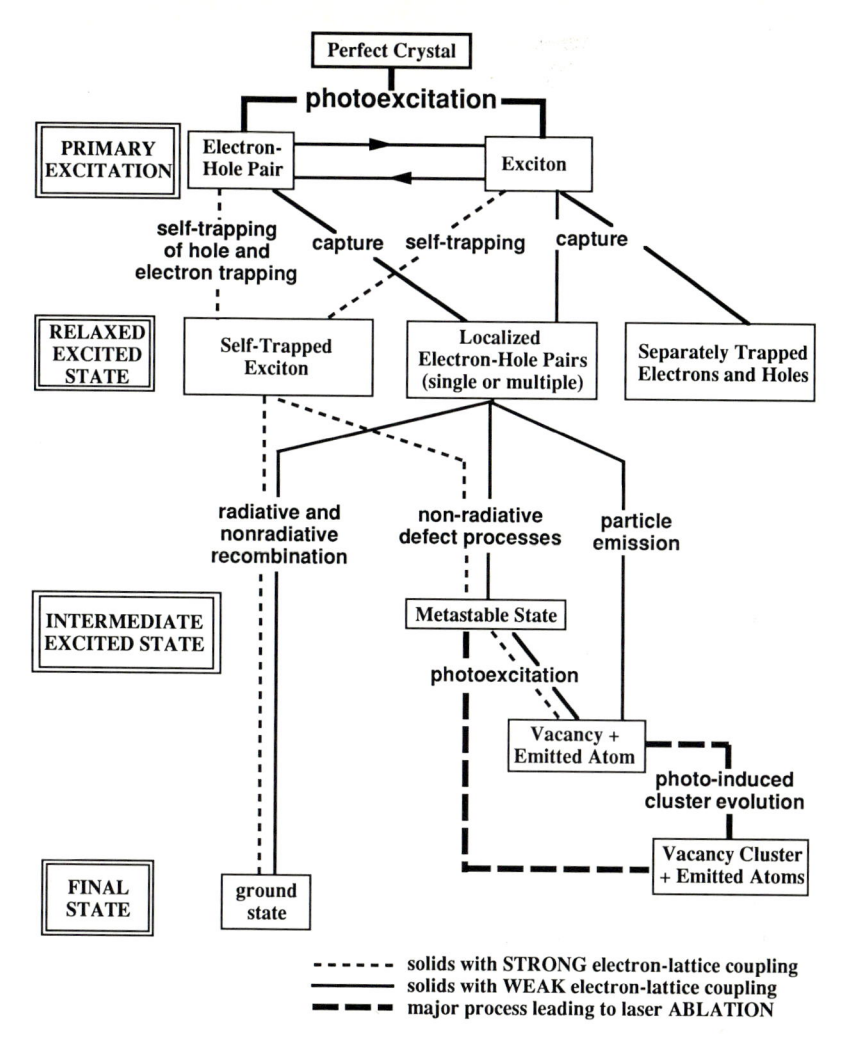

Fig. 2.3. Schematic of the sequence of events leading to formation of the relaxed excited state, the intermediate excited state, laser-induced desorption and, after formation of vacancy clusters, laser ablation

may produce rapid local heating; thermalization times required to bring the energetic electron to the bottom of the conduction band are of the order of 1 ps.

To develop a kinetic model of ablation for semiconductors, we are especially interested in the total scattering cross section. Theoretical treatment of the scattering may be chosen based on the electron energy relative to the size of the scattering potential; for slow electrons, effective-range theory is appropriate, while for fast electrons, one may reasonably use the Born approximation. The details of scattering will depend on whether the mechanism is a resonant, charged-particle or a neutral-defect interaction. The range of typical cross sections may be found in [2.15].

2.3.2 Capture of Charge Carriers at Defect Sites

In an electron–hole plasma of concentration n, each defect acts as a recombination center with a characteristic lifetime of $1/(\sigma_t v n)$, where σ_t is the carrier trapping cross section, v the velocity of the charge carrier and n the concentration. With $n = 10^{19}$ cm^{-3}, the time constant can be as short as 1 ns.

Following the initial photoexcitation, an electron or a hole is first captured into a high-lying excited state; relaxation follows through a series of non-radiative transitions among highly excited states. Further details of non-radiative de-excitation processes are described in Sect. 2.4. Here, we emphasize two points. First, the capture cross section is governed by the rate of de-excitation to lower levels, since delocalization of trapped carriers from highly excited states is very probable. Capture is, in fact, the outcome of a competition between the delocalization and the transition to low-lying excited states. Second, de-excitation to the lowest excited state is usually much faster than the transition from the lowest excited state to the ground state. At the lowest excited state the lattice is relaxed to a configuration in which the Adiabatic Potential Energy Surface (APES) has a minimum, as shown schematically in Fig. 2.4. From this point, the detailed dynamical evolution of the system depends on the strength of the electron–lattice coupling.

Successive capture of an electron and a hole (or vice versa) leads to recombination of an electron–hole pair. Although the process is not a major loss channel for excitation energy when defect concentration is low and excitation density is extremely high, it is during the process of recombination that the electronic energy is imparted to local modes which ultimately lead to particle emission. Just as in the case of charge-carrier capture, the relaxed excited defect state will be formed after successive capture of both carriers. The energy possessed by an electron–hole pair is partially stored in this excited state. We shall see later that the recombination channel can be the source of desorption and even ablation.

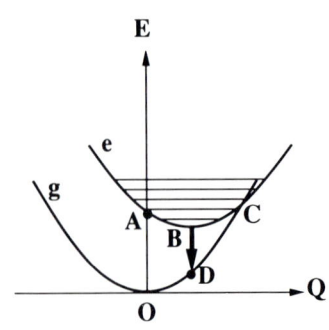

Fig. 2.4. Competition between radiative and non-radiative transitions from the excited-state APES to the ground state

2.3.3 Lattice-Induced Localization of Free Carriers and Excitons

For laser-induced desorption and ablation, the most important electron–lattice interactions are those which lead to localization of the energy stored in the electron–hole pair created by photon absorption. This is because the electronic energy needs to be imparted to local modes for any lattice rearrangement to take place. The first stage of this process is the electron–lattice interaction which leads to the *localized excited state* – the precursor to the *intermediate excited state* considered in the following section.

In general, an electron, hole or exciton created by photoexcitation of a solid can reduce its energy by delocalizing in a band state or localizing through lattice polarization. The criterion for localization is determined by the relative values of the bandwidth 2B, and the lattice relaxation energy E_{LR}. Self-trapping occurs if $B < E_{LR}$ [2.28, 29]; solids for which this is the case are said to have strong electron–lattice coupling. Weak electron–lattice coupling is characteristic of solids for which $B > E_{LR}$. If more than one carrier may be localized at a single site, it is necessary to incorporate into the localization criterion the on-site Coulomb repulsion energy U. *Toyozawa* [2.30] developed the phase diagram in coordinates B, E_{LR} and U shown in Fig. 2.5. If E_{LR} dominates over B and U, two-carrier localization can occur; if U and E_{LR} dominate B, localization of a carrier results. Alternatively, if B dominates the others, the carriers remain free. The criterion for two-carrier localization on defects ($E_{LR} > U$) was suggested first by *Anderson* [2.31].

In solids with weak electron–lattice coupling, luminescence competes effectively with the motion of free carriers. In solids with strong electron–lattice coupling ($E_{LR} > B$), on the other hand, free carriers can be trapped by their self-interaction with the lattice distortion field. Self-trapping of a charge carrier can be expressed in terms of an APES similar to that shown in Fig. 2.6b; the minimum of the APES determines the configuration of the self-trapped carrier. There is also an illuminating chemical interpretation of self-trapping in the case of holes: the self-trapped hole is localized because it is bound in a covalent configuration in which it occupies the highest anti-bonding orbital of the combined species [2.32]. In many insulating solids, self-trapped holes are

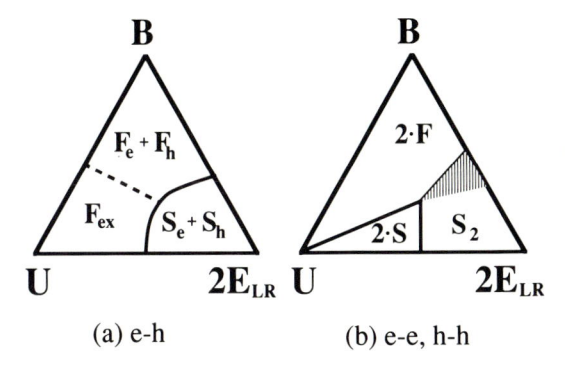

Fig. 2.5. Phase diagram for localization of two carriers of (a) unlike sign (*e–h* pair) and (b) like sign (*e–e* or *h–h* pairs). The terms U, B, and E_{LR} refer to the Coulomb interaction energy between carriers, bandwidth and lattice-relaxation energies, respectively. The letters "F" and "S" refer to free and self-trapped carrier pairs, respectively

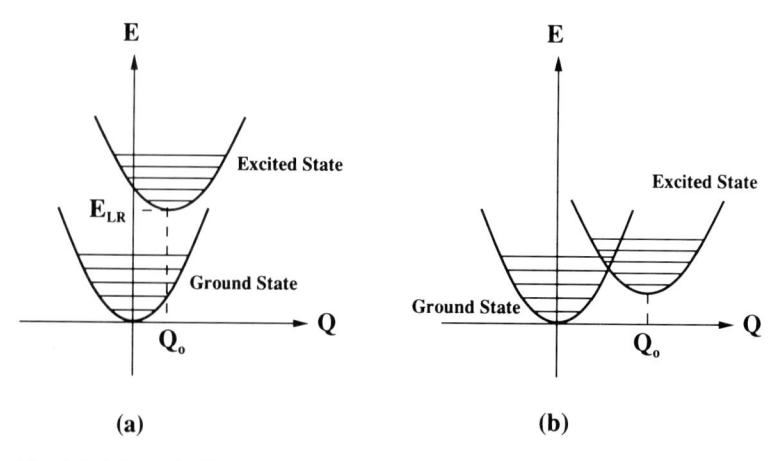

Fig. 2.6. Schematic illustration of two possible adiabatic potential energy surfaces. (**a**) The case where the excited-state APES minimum lies above the ground-state APES. (**b**) The case in which the APES minimum of the excited state lies below the ground-state APES at some point

precursors of the self-trapped exciton which has significant photochemical consequences for the evolution of the material under laser irradiation.

A Self-Trapped Exciton (STE) is formed when a self-trapped hole traps an electron or as the result of free-exciton relaxation. The STE can be given a chemical interpretation analogous to the self-trapped hole, but the extra electronic excitation energy of STEs makes possible the formation of a variety of relaxed states [2.33]; a Frenkel pair is one possible relaxed state, for example [2.34]. The energy gain by self-trapping of an exciton is larger than that for a hole [2.35]. It is believed that STEs may play an important role in laser-induced desorption in the alkali halides, alkaline-earth fluorides and amorphous SiO_2.

Two-hole localization by the Anderson negative-U interaction on a defect site has been observed in the bulk of semiconductors [2.36, 37]. For defects on surfaces, one possibility for localization of two holes is that in which an atom on the surface is displaced out of the surface [2.38]. In this case, Anderson localization eventually leads to desorption. A potential barrier due to Coulomb repulsion may exist for two-hole localization in a dense electron–hole plasma. We point out later that a two-hole localized state can be produced not only by trapping of two carriers, but also by cascade excitation of a defect site. Defect-mediated two-hole localization occurs in weak-coupling solids, where it is initiated by trapping of carriers.

In summary, localization of electronic excitation energy occurs via self-trapping in strong-coupling solids, while defects are the only available source for localization in weak-coupling solids. In the former, the electronic excitation energy is localized partially by self-trapping of an exciton or by trapping of an electron by a self-trapped hole to form a self-trapped exciton. In the latter case, successive trapping of an electron and a hole (or vice versa) at a defect in an excited state localizes the electronic excitation energy. So far, competition between self-trapping and two-carrier localization which may appear in strong-coupling solids (Fig. 2.5, diagram for F_h and F_e) has not been observed.

2.4 Creation and De-Excitation of the Localized Excited State

Once charge carriers or excitons have been localized, the lattice relaxes to a configuration at a minimum of the APES on a time scale less than the inverse of the characteristic lattice frequency. It is convenient to distinguish two types of relaxed configuration as shown in Fig. 2.6: (a) with the minimum of the relaxed excited state potential surface located above the ground state, and (b) with the minimum of the relaxed excited state potential-energy surface located below the ground state APES. In case (a), both radiative and non-radiative transitions lead to the ground state. Since the energy separation between the relaxed excited state and the ground state is large, the non-radiative transition involves a multiphonon process which in turn requires strong electron–lattice coupling. In case (b), because of the large lattice relaxation energy E_{LR} and the configuration change at the APES minimum, the electronic excitation energy is largely transferred to configurational energy. Electronic excitation leads to a metastable state from which the recovery to the ground state proceeds via a thermally activated process. Gas-phase photodissociation of diatomic molecules and the analogous process for atomic desorption of chemisorbed species from metal surfaces belongs to this category; the relaxed configuration is the one from which an atom is ejected.

Radiative recombination, which generally has a lifetime ~ 1 ns, cannot be the major channel of energy dissipation. Thus, we concern ourselves mainly with non-radiative transitions involving the following elementary processes:

1) Conversion of electronic energy to vibrational energy;
2) Conversion of electronic to configurational energy;
3) Transfer of vibrational energy from one degree of freedom to another;
4) Transfer of electronic energy from one configuration coordinate to another.

These processes have been reviewed by several authors [2.13, 39, 40]. Most treatments of non-radiative transitions in laser–solid interactions have direct relevance to laser–semiconductor interactions, and reviews of this literature already exist [2.41, 42]. In strong-coupling solids, there is much relevant work on defect transitions in the alkali halides, and to a lesser degree for the fluorides and some oxides [2.43].

2.4.1 Non-Radiative De-Excitation

Non-radiative de-excitation from a relaxed excited state located above the ground state occurs either through conversion of electronic energy to vibrational energy (process 1) or by a transition to a vibrationally excited electronic ground state (process 2), followed by cooling transitions (process 3) to a state still lower in total energy.

Referring to Fig. 2.4, we can distinguish two types of non-radiative transitions from the excited state e to the ground state g: one after the cooling transition to the lowest vibrational state on e, or from the relaxed state B, and the other from vibrationally excited states of e. The former are anticorrelated with the temperature dependence of the radiative transition from B to the ground state D. The transition probability depends critically on the energy difference between B and the crossing point C of the ground and excited states through the Pekarian factor [2.44]. The transition probability includes both thermally-activated and temperature independent parts and, since it is competing with radiative transitions, is of the order of the radiative transition probability at low temperatures, increasing rapidly with temperature.

Non-radiative transitions from vibrationally excited states of e are sometimes called *Dexter-Klick-Russell* (DKR) processes [2.45]. Suppose that a photon is absorbed by a system in the ground state at an equilibrium configuration O. Since no atomic motion is induced during photoexcitation, the transition is a vertical or a Franck-Condon transition, producing an excited state at configuration O. Immediately after excitation a phonon wave packet is located at A, which oscillates following the APES of the excited state. If the state A lies above the crossing point C, a non-radiative transition to the ground state has a high probability when the packet reaches C, which occurs on a time scale less than the inverse of the characteristic lattice frequency. DKR processes are especially effective in the process of de-excitation following capture of a charge carrier in a higher excited state, since, if the level crossing between the two excited states is above point C, the subsequent transition to the lower state is fast.

Radiative recombination removing electronic energy from the surface is strongly favored if the crossing point is located above the point A on the excited-state potential surface to which the transition is made. The transition probability during cooling transition from the excited state is enhanced in this condition. This suggests that the rate of the non-radiative transition to the ground state depends on the exciting photon energy. It may also explain recent observations of the rapid non-radiative transition from electron–hole pairs to the lowest excited state of the self-trapped exciton [2.46].

The transition from e to a vibrationally excited state g is followed by cooling transitions as vibrational energy is transferred to other local modes. Vibrational energy is transferred to those local modes strongly overlapping with those of the state g, and then converted to delocalized lattice phonon modes at a slower rate. Of particular interest is the case when the vibrational energy is transferred to a reaction mode Q_R that causes local configurational change or desorption, as shown schematically in Fig. 2.7. Here the vibrational energy in the Q mode surpasses the potential barrier for the reaction. This process is often called *phonon kicking* and has been suggested as the cause of recombination-induced defect migration [2.47] and of desorption from rare-gas solids [2.48]. If a

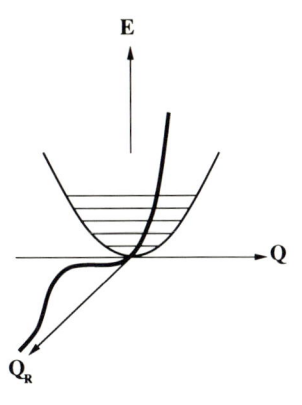

Fig. 2.7. Illustration of the APES appropriate for the direct transfer of electronic energy in the configuration Q to the reaction mode Q_R

configurational change fails to occur, the localized vibrational energy is ultimately converted to lattice phonons, that is, to heat.

2.4.2 Transfer of Electronic to Configurational Energy

When the APES of an excited state has a minimum at a configuration substantially different from that for the minimum of the APES of the ground state, as depicted in Fig. 2.6b, excitation results in configurational change [process (3)]. Unlike phonon-kicking, where the electronic energy first goes to the vibrational energy of a local mode and then to the reaction mode, the electronic energy goes directly to the reaction mode in this case. Desorption induced by excitation of a substrate–adsorbate system to an antibonding excited state, the Menzel-Gomer-Redhead mechanism, is one example of this process in photo-induced desorption.

Formation of a metastable excited state by electronic excitation of defects – as in Fig. 2.8 – in semiconductors is another well-documented example of this type of energy transfer [2.49]. For example, when an antisite defect in compound semiconductors – for example, an As atom occupying a Ga site of GaAs – is electronically excited, the defect is converted to a metastable state with

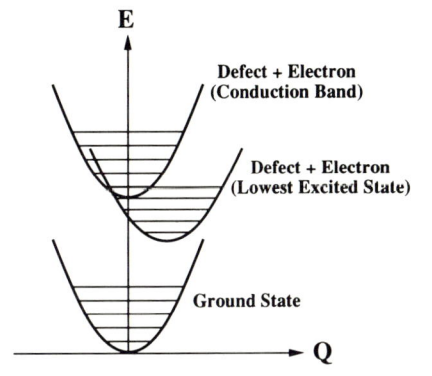

Fig. 2.8. Schematic illustration of the APES for defect states in semiconductors

a configuration significantly different from the ground state [2.50]. Configurational change by electronic excitation may be regarded as the conformation of the lattice to minimize the total energy; the lattice after excitation is more flexible because the bonding strength has been weakened by electronic excitation [2.51]. The concept is an extension of the earlier suggestion by *Van Vechten* et al. [2.52] that the formation of holes in the valence band weakens the bonds and reduces the activation energy for defect migration. Localized holes are especially effective in bond weakening; the excitation of an electron from a bonding to an anti-bonding state in a tetrahedrally coordinated semiconductor softens the dispersion curve of the transverse acoustic phonons, as pointed out by *Wautelet* and *Laude* [2.53].

The best understood example of the conversion of electronic energy to vibrational and configurational energies is the STE and its related defects in alkali halides [2.32, 43]. The STE in alkali halides consists of a halogen molecular ion X_2^- in an off-center configuration, and an electron; it is the precursor of the metastable state consisting of an F-center and an H-center (Frenkel pair on the halogen sublattice) [2.34]. The initial optical excitation forms an electron–hole pair, which in turn leads, first by self-trapping and then through a series of non-radiative and cooling transitions, to the STE. From the STE, the Frenkel pair may be generated either by a simple thermal hopping over the barrier [2.54, 55, 56], or by a dynamical transfer from the STE to the Frenkel-pair configuration via a series of cooling transitions on the APES for the lowest state of the STE. This latter possibility has now been confirmed experimentally using ultrafast laser techniques [2.46].

2.4.3 Other Non-Radiative De-Excitation Channels

Non-radiative de-excitation can also occur by conversion of electronic energy to other electronic degrees of freedom (process 4), if this second electronic system transfers its energy to vibrational modes or configurations. The Auger transition is an important example of this process: The localized electronic excitation energy is transferred to free electrons which in turn are de-excited by emitting phonons [2.57]. This process dominates when the density of electron–hole pairs is high, as it can be in semiconductors. However, it also leads to heating of the lattice and hence is irrelevant to the electronic processes of desorption and ablation unless the temperature rise is extremely high.

2.5 Survey of Experimental Results

Pulsed-laser annealing of Si was originally ascribed to thermal mechanisms [2.58]. Since then, many laser-induced desorption and ablation experiments have clearly shown the importance of electronic rather than thermal processes.

2. Electronic Processes in Laser Ablation of Semiconductors and Insulators 27

Table 2.2 Electronic Processes Exhibited in Laser Desorption and Ablation

Electronic Process in Desorption/Ablation	Material(s) Showing Effect
Transition from linear to nonlinear yield vs excitation density	$LiNbO_3$, $KNbO_3$
Nonthermal velocity distributions of emitted atoms and ions	CaF_2, BaF_2, NaCl
Molecular emission, nonthermal internal energy distribution	Al_2O_3
Multiphoton desorption enhanced by electronic resonances	BaF_2, CaF_2
Change in desorption yield induced by defect formation	GaP, GaAs
Enhanced ablation yield induced by prior defect formation	GaP, GaAs
Molecular cluster emission varying as function of wavelength	$LiNbO_3$, $KNbO_3$
Desorption induced by selective excitation of surface states	BaF_2, GaP, GaAs
Variation in ablation threshold with electronic structure	Semiconductors

Indeed, it appears that Si may have been the exception which proves the rule. Here we review experimental evidence for electronic processes in laser-induced desorption and laser ablation from halides, oxides and compound semiconductors, summarized in Table 2.2. This sequence corresponds to strong, intermediate and weak electron–lattice coupling, respectively. We make no attempt to separate the two phenomena here; the experimental distinction between desorption and ablation is not always clear-cut.

2.5.1 Alkali Halides and Alkaline-Earth Fluorides

In alkali halides and alkaline-earth fluorides, ground-state atoms and ions are desorbed with a linear yield–fluence relation at low density of excitation by band-to-band transitions produced by conventional ultraviolet sources [2.59], by multiphoton excitation [2.60], and by electron-beam irradiation [2.61]. *Syzmonski* et al. [2.62] have measured velocity distributions of halogen atoms desorbed by electron irradiation, and found both thermal and hyperthermal components. Furthermore, the fast atoms are emitted along the ⟨100⟩ direction perpendicular to the ⟨100⟩ surface, rather than along the ⟨110⟩ direction. Although the details of the desorption mechanism remain controversial, it is known that an exciton or an electron–hole pair is transformed to a Frenkel pair in the halogen sublattice. It is also generally accepted that halogen-atom emission is the primary process. In the linear regime, the surface is decomposed progressively, so that a nonstoichiometric-damaged surface results from irradiation. Thermal emission of alkali atoms from the alkali-rich surface follows. This picture of halogen emission as the primary processes has recently been confirmed in measurements and model calculations of halogen-atom desorption from alkali-halide clusters [2.63].

Almost all laser-ablation experiments on alkali halides and alkaline-earth fluorides have been carried out using laser beams of sub-gap photon energies. The yields of halogen and alkali atoms vary linearly with the density of excitation at low fluence, becoming superlinear at higher fluences, as illustrated by the data of Fig. 2.9 [2.64]. In the superlinear regime, the yield–fluence relation

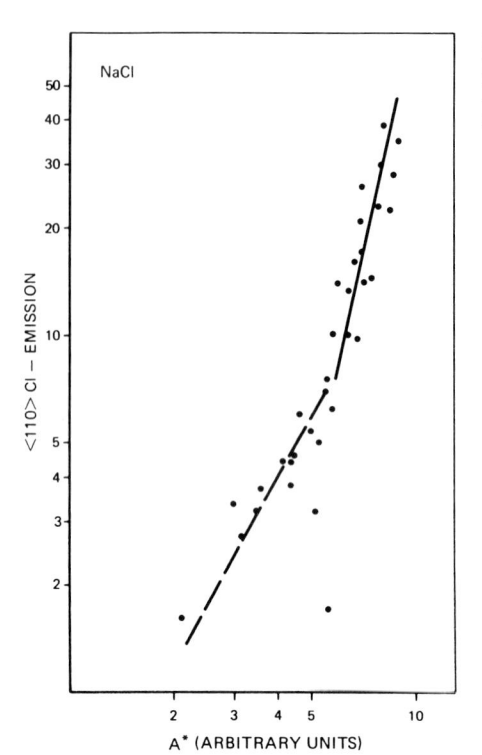

Fig. 2.9. Yield of Cl ions from NaCl under laser irradiation by the second harmonic of Nd:YAG (0.532 μm) showing the linear and nonlinear regimes

for emission of atoms obeys a power law in the density of excitation with exponents of 2–3. As the laser fluence increases, ions and electrons are emitted and a laser-induced plasma eventually forms. In general, the ignition of the surface plasma marks the transition from desorption to the ablation regime and occurs over a wide range of materials, pulse lengths and wavelengths near an intensity threshold of $\sim 2 \times 10^8 - 10^9$ W cm^{-2} [2.65].

The mechanism which produces the superlinear yield–fluence relation is widely believed to be related to the formation of defects. Defects contribute to the superlinear ablation yeild either by providing a real intermediate state to enhance the multiphoton transition probability [2.66] or by weakening the bonds of neighboring atoms to the point where these can be ejected by one-photon excitation. The observed differences between alkali halides and alkaline earth fluorides with respect to the accumulation of metal layers under laser irradiation probably arises from differences in vapor pressure of the metallic component.

2.5.2 Oxides

Oxide materials are critical to optical, electronic and superconducting technologies. Laser ablation is already in use to drill vias in SiO_2 layers in microelectronic circuits, for thin-film deposition of high-T_c superconductors and for machining optical materials and ceramics. Oxides have electron–lattice coupling strengths intermediate between the halides and the semiconductors; laser-induced particle emission and ablation are therefore expected to show behavior intermediate between strong- and weak-coupling solids. Recent experiments show that laser ablation in oxides is often electronic or strongly influenced by electronic mechanisms.

Laser ablation has been studied extensively in sapphire (c-Al_2O_3). Whether excitons are self-trapped in this material remains controversial [2.67]. *Dreyfus* and co-workers [2.68] carried out steady-state ablation experiments on sapphire with a 248-nm KrF laser, using laser-induced fluorescence to detect both Al atoms and AlO molecules. Above the ablation threshold (0.6 J cm^{-2}) the velocity distributions of the AlO molecules were inconsistent with surface melting, suggesting an electronic mechanism for ablation [2.68]. Also, photothermal deflection measurements showed a temperature rise insufficient to account for the kinetic energies of the ablated AlO and Al [2.69]. AlO ablated from a thin sapphire film on an aluminum substrate had similar velocity distributions even though the estimated temperature change in the film was significantly different from that of the bulk sample, suggesting a common electronic mechanism [2.70]. *Chase* has reported that laser ablation of Al_2O_3 at the fundamental wavelength of the Nd:YAG laser ($hv = 1.17$ eV) induced a surface reconstruction accompanied by emission of Al$^+$ ions with a typical energy of 8 eV [2.71], possibly from the decay of an exciton [2.72].

Oxide glasses have great practical importance in many technologies. Thus it is surprising that so few laser-desorption experiments have been carried out on these materials. Amorphous and crystalline SiO_2 support the self-trapping of excitons, but only in a-SiO_2 do the STEs relax to form Frenkel pairs [2.73]. In fused silica, the E' and non-bridging oxygen-hole centers are the analog of the F- and H-centers in the alkali halides. *Tsai* et al. [2.74] have shown that 6.4 eV photons from an ArF laser can produce the E' point defect in fused silica by the nonradiative decay of neutral excitons. *Devine* [2.75] has reported that the density of E' defects in a-SiO_2 scales with the number of two-photon transitions at 6.4 eV. The creation of optically absorbing defects in fused silica by two-photon absorption at 193 nm has also been observed [2.76]. This suggests that defect creation should have a significant role in laser-induced desorption and ablation from these solids. Indeed, recent studies in silicate glasses show that prior irradiation by electrons to create electronic defects dramatically lowers the laser ablation threshold [2.77].

Laser ablation of MgO has recently been studied because the crystal has a wide bandgap and is very ionic, but band-to-band excitation does not produce electronic defects. It appears that self-trapping of excitons in MgO is induced above liquid-nitrogen temperature but not at very low temperatures [2.78]. Studies of laser ablation at 248 nm in neutron-irradiated MgO shows that the introduction of displacement defects dramatically lowers the fluence threshold for laser ablation [2.79], Even above the plasma threshold ($\Phi > 2$–5 J cm^{-2}), incubation effects consistent with defect accumulation were an important factor in the rate of material removal. Interestingly, laser irradiation at high fluence is also found to clean up or remove some surface or defects, suggesting that laser processing can remove material and stabilize surfaces. Similar surface-defect removal has been found in laser-induced desorption from weak-coupling solids.

Laser ablation from ternary oxide ferroelectrics is important in nonlinear optics applications, micromachining and direct writing in optical circuits, and in thin-film deposition by laser ablation [2.80]. Because the bulk conduction-band edge for oxide ferroelectrics lies 3–4 eV above the valence-band edge, it is possible to use visible and ultraviolet lasers to study desorption and ablation both at low and high excitation densities, depending on the relationship between hv and E_{gap}. Ultraviolet laser irradiation above threshold near the band edge produces a plethora of ground-state atoms, ions, diatomics and excited atoms [2.81]. Below the ablation threshold, one observes efficient desorption of molecules and molecular ions [2.82].

The current demand for high-quality thin films of high-temperature superconducting materials has produced a rich outpouring of literature on the relationship of ablation parameters to thin-film quality, but rather less work on fundamental mechanisms. One of the reasons for this is that high-quality film growth takes place primarily in an oxygen-rich atmosphere at moderate pressures. This gaseous atmosphere produces many complicated secondary effects, such as plasma formation, due to the interaction of the dense, stagnating layer of ejecta from the surface with the incoming laser light. However, thus far the literature on desorption and ablation in ultrahigh vacuum, where one might hope to observe the primary processes, is rather sparse [2.83].

Finally, at the weak-coupling limit of binary oxides, we note that laser ablation of ZnO shows the existence of a laser ablation threshold and suprathermal velocity distributions of the ablated atoms for one-photon band-to-band transitions induced by a laser [2.84].

2.5.3 Compound Semiconductors

Laser ablation of semiconductors has been reviewed by *Nakai* et al. [2.85]. Most early measurements of laser-induced desorption from compound semiconductors used conventional mass spectrometers. Because of the low sensitivity of these devices, several layers of atoms per laser pulse had to be ablated in order to reach acceptable signal levels, so that these data clearly stem from the ablation regime. The ablation threshold of both elemental and compound semiconductors was measured by *Ichige* et al. [2.86] and shown to decrease with increasing ionicity (Fig. 2.10). In view of the close correlation between ionicity and bond strength [2.87], the data clearly suggest, not surprisingly, that the bond strength plays a role in ablation.

A more sensitive detection technique, resonance ionization spectroscopy, has been employed recently by the *Nagoya* group to study the mechanisms of laser-induced particle emission as well as ablation. A typical example of the yield–fluence relation for Ga^0 from GaP at fluences near the ablation

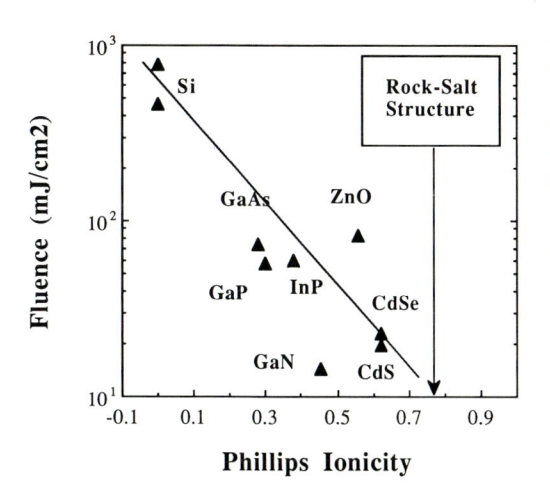

Fig. 2.10. Decrease of laser-ablation threshold with increasing ionicity, showing the highest fluence thresholds for the Group IV elemental semiconductors, the lowest for the II–VI compound semiconductors. The arrow indicates the transition ionicity for forming the tetrahedral semiconductor bond

threshold is shown in Fig. 2.11 [2.88]. The yield at threshold increases more rapidly with laser fluence in compound semiconductors than in alkali halides. The ablation threshold was determined not only from the fluence dependence of the yield but from observation of the modification of the LEED pattern [2.89]. The extent of damage to the ablated surface was monitored by changes in stoichiometry deduced from Scanning Electron Microscopy (SEM), and from Auger Electron Spectroscopy (AES). Photoelectron emission has been shown to exhibit nearly the same dependence on the fluence as the ion emission yields [2.90]. Both ions and neutral atoms are emitted by laser irradiation, the yield of the former increases more rapidly as the fluence increases [2.91]. In some experiments, the surface of GaP was bombarded with a single shot, 100 shots and 1000 shots with increasing fluences, and the ablation threshold was determined by observing changes in the LEED patterns. The ablation threshold

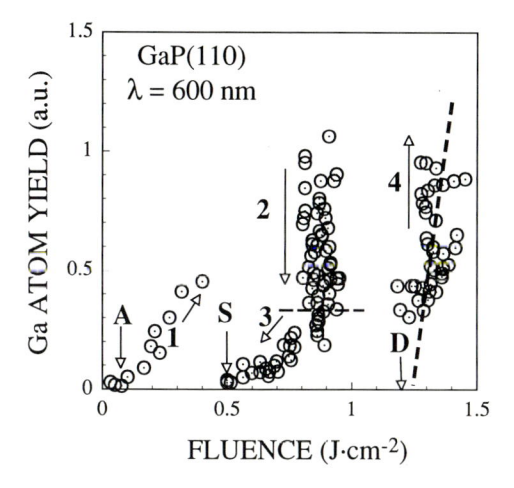

Fig. 2.11. Near-threshold measurement of Ga^0 yield from GaP (1 1 0) for laser excitation of an unfilled surface state, illustrating the existence of variations in threshold depending on the type of site

was found to depend only slightly on the number of laser pulses, indicating that the ablation is not due to accumulation of damage from successive laser pulses [2.85].

High-sensitivity measurements on GaP and GaAs have revealed that laser-induced particle emission occurs at fluences much lower than the ablation threshold [2.23, 92]. The emission yield in this fluence range is enhanced by Ar^+ irradiation and reduced by annealing, indicating that the emission is defect-related. Three distinctive types of yield–fluence relations are observed with thresholds corresponding to the intensities labeled I_A, I_S and I_D in Fig. 2.11. The component of the yield enhanced most noticeably by introduction of defects by Ar^+ irradiation is reduced by repeated irradiation with tens of laser pulses at the same fluence. In addition to this defect type, however, measurements of emission yield as a function of laser shot-number reveal another defect type which is eliminated slowly (hundreds to thousands of pulses) upon repetitive nano-second-pulse laser irradiation. The former is ascribed to adatoms and the latter to kink sites on steps. Yields of these components are lower for laser photons above the bandgap energy than for below the bandgap energy and interpreted as that the electron–hole pairs confined in the surface states are particularly efficient in causing the emission [2.19, 92, 93]. The third distinctive type of defect, from which ablation is initiated at fluences above 1 J cm^{-2}, *increases* with repeated laser irradiation [2.94].

Another study in the III–V semiconductors by *Long* et al. [2.95] has shown that surface decomposition occurs in GaAs by a 5 ns high repetition rate (6 kHz) copper-vapor laser ($\lambda = 510$ nm) irradiation at fluences as low as 1 mJ/cm^2. The electronic character of the process is assured, since the temperature rise due to photon absorption is less than 13 K. Time-resolved photoemission measurements show the growth of a Fermi-level feature related to the formation of Ga islands, indicating surface decomposition. Scanning electron microscope measurements show the Ga islands to have diameters as large as 100 Å. This measurement differs from desorption measurements by the *Nagoya* group in two significant respects: the photon energy used is far above the bandgap energy, and the pulse-repetition frequency is very high, possibly inducing long-term relaxation effects.

The II–VI compound semiconductors are of increasing technological interest and have begun to attract attention in studies of laser ablation, primarily for the purpose of modifying surface epitaxy. Recent experiments by *Brewer* and coworkers have shown that CdTe surface composition can be modified reversibly by excimer (KrF, $\lambda = 248$ nm) laser irradiation [2.96]. The experiments were carried out in high vacuum, and the removal rates ranged from less than 1 Å/pulse to approximately 100 Å/pulse. Above a fluence threshold of 40 mJ/cm^2, a metal-enriched surface is observed; below this threshold, stoichiometry was restored by laser irradiation. The velocity distributions of the Cd, Te and Te_2 desorption products are all thermal in this regime. The experimenters explain their results as the competition between two thermally activated processes, under the assumption that the kinetics of Cd desorption is linked to the kinetics of Te_2 formation. This is a question which revolves around detailed information on the chemical kinetics which is fundamentally an electronic process.

2.6 Models of Laser-Induced Desorption

Laser-induced desorption experiments show that in all solids with a finite bandgap, there is a succession of electronic processes which is remarkably similar in all cases: production of electron–hole pairs, followed by lattice-localized relaxation, a transition to a relaxed excited state and, finally, a transition to an antibonding potential energy surface on which one or more species move away from the surface. Our objective in this section is first, to summarize a number of models which have been proposed to explain laser-induced particle emission and, second, to show, using selected examples, how these models can be developed in a quantitative way to produce detailed dynamical information about the process of laser-induced desorption in a variety of materials.

2.6.1 Models for Electronic Processes in Laser-Induced Desorption

Several models of electronic processes are described in the literature which attempt to explain Laser-Induced Desorption (LID) at fluences below the ablation threshold. Most are based on the conceptual picture of the *Menzel-Gomer-Redhead* (MGR) model [2.97] of molecular photodissociation at surfaces. Although this model with its picture of a Franck-Condon transition to an excited, anti-bonding electronic state is too simple for detailed guidance, it is a useful starting point. What must be added to the MGR model is the material-specific character of the anti-bonding electronic state which leads to emission of an atom. Since the fundamental photon–solid interaction produces electron–hole pairs or excitons with delocalized wave functions, whereas the formation of the MGR anti-bonding state requires localization, the fundamental issues are the kinetics and dynamics of localization and the anti-bonding nature of the localized state.

In solids with strong electron–lattice coupling, self-trapping of excitons and holes provides a mean of localization. Furthermore, the self-trapped excitons in the bulk are known to be converted to a variety of configurations including close interstitial-vacancy pairs. Not much information on self-trapping near surfaces has been obtained, except that conversion from a self-trapped excitons to an emitted atom is energetically feasible. Figure 2.12 depicts possible configurational changes induced after formation of an exciton en route to emission of an atom [2.98]. Evidently, several modes of configurational change are involved: first, the halogen–halogen distance Q_1 to form a halogen molecular ion X_2^-; second, the translational motion Q_2 of X_2^- toward the surface; and third, the

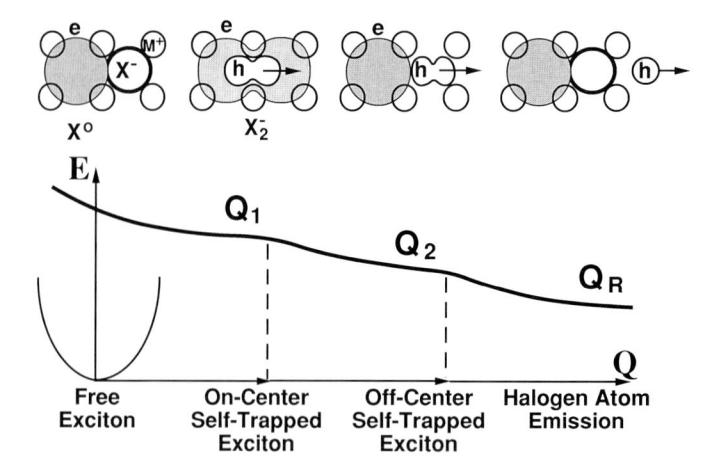

Fig. 2.12. Progression of steps to laser-induced desorption in a strong-coupling solid, showing the series of configuration changes from the initial electron–hole pair to the self-trapped exciton (relaxed excited state); the F–H center pair (intermediate excited state); and the ultimate ejection of the halogen atom from the surface

decomposition mode Q_3 of X_2^- near the surface. We note that the APES as a whole has a MGR-type anti-bonding nature, although it involves several configurational coordinates. The APES describes emission of halogen atoms leaving halogen vacancies, resulting in a change of stoichiometry [2.99].

The excitonic mechanism described above is considered to emit halogen atoms along the $\langle 110 \rangle$ directions. Based on experimental evidence that the emission is directed along a $\langle 100 \rangle$ trajectory, *Szymonski* has suggested that when a "hot hole" approaches the surface with finite kinetic energy, the kinetic energy is converted to that of the emitted atom [2.100]. However, recent quantum mechanical calculations of the adiabatic potential energy for the decomposition of an exciton near the surface [2.101] predict that the decomposition of an exciton leads to emission along the $\langle 100 \rangle$ direction. Until further experimental evidence is accumulated, we take the excitonic mechanism to be the cause of emission, since the excitonic mechanism, in which the internal energy of an exciton can be converted to kinetic energy in the emitted atom, is energetically favorable. Emission of alkali atoms will be induced from alkali-rich surfaces both by thermal evaporation and by electronic excitation. Emission of excited alkali atoms by single-photon transitions from the valence band has been observed on alkali-rich surfaces, although the mechanism is not yet clear [2.102].

A similar configuration coordinate diagram describes laser-induced desorption in alkaline earth fluorides, in which the relaxation of excitons to self-trapped excitons and F-H pairs is analogous to that in alkali halides. Although the existence of self-trapped excitons in SiO_2 is well established, the fate of self-trapped excitons in its clean surface is not yet clear. Since self-trapping arises partly from weakening the bond on which a hole is localized, it is conceivable that an exciton near the surface leads to emission of an atom. Relaxation of excitons near the surface of crystalline and amorphous SiO_2 is expected to resemble that in the bulk solid [2.103].

In solids with weak electron–lattice coupling, neither holes nor excitons are self-trapped and thus another mechanism of energy localization must be operative to produce laser-induced desorption. The two-hole localization model of *Itoh* and *Nakayama* [2.38, 104] was first proposed to explain the laser sputtering of Zn atoms from ZnO which showed ablation only above a threshold value of the fluence and a two-temperature Maxwell-Boltzmann velocity distribution. The two-component velocity distribution shows that one is not dealing with an equilibrium process, such as melting. *Itoh* and *Nakayama* proposed a phonon-assisted localization of two holes at a single surface site based on *Anderson*'s idea that the Coulomb repulsion might be less than the energy gain from an accompanying lattice distortion. The threshold phenomenon was explained, in turn, by noting that a certain density of the electron–hole plasma is required to screen the effective charge of the two holes until they can move within the effective interaction distance of the Anderson negative-U potential. The screened Coulomb potential is given by

$$U_{sc} = \frac{e^2}{\varepsilon(n)r_c} \exp[-2k(n)r_c], \tag{2.1}$$

where n is the density of the electron–hole plasma, $\varepsilon(n)$ is the dielectric constant and r_c is the screening radius which is of the order of one or two lattice constants in a dense e–h plasma. The probability for particle emission can be calculated from reasonable assumptions about the density of the laser-induced electron–hole plasma.

A more general approach to this concept has been proposed by *Toyozawa* [2.29] for localization of two electrons or holes in the bulk. *Sumi* [2.105] has recently proposed an alternative to the idea of localization induced by a high-density electron–hole plasma which screens out Coulomb repulsion. Instead, he proposed that the Coulomb barrier to two-hole localization is overcome if the density of electron–hole pairs is sufficient for the Fermi energy of the degenerate holes to exceed the potential barrier. The models described in this section assume that the desorption of atoms from semiconductors occurs at perfect surface sites; the negative-U interaction is strong enough to weaken the bonds for the atom. So far as described in the last section, laser ablation in the weak-coupling limit is more likely to be defect related. Thus judgement on these models should be reserved until further experimental evidence is accumulated.

In solids with weak electron–lattice coupling, defects can also localize electronic excitation energy and several defect-related desorption models have been proposed. For example, *Wu* [2.106] proposed that Si^+ and Ge^+ are emitted by Coulomb repulsion due to two-hole localization induced by a defect-associated e–hh Auger process. The energy of the holes which makes the Auger transition possible is supplied by energetic holes; thus, the emission yield is considered to depend on the incident photon energy, but not on the laser intensity except to the extent that the electron–hole plasma near the surface is instrumental in weakening the surface bonds. By taking into account the differing optical properties of the Si and Ge surfaces and using the Phillips bond-charge model for computing the bond-weakening produced by the energetic hole, he finds a ratio for Si^+ to Ge^+ in reasonable agreement with experiment.

Hattori et al. [2.27, 107] have proposed a phonon-assisted multi-hole localization mechanism to explain the superlinear dependence of the yield. In this model, an excited defect state is created either by sequential trapping of an electron–hole pair or by defect excitation. The defect-excited state is relaxed to form a metastable state, which is excited again, during the same laser pulse, to form a different excited state. In each cycle, a new hole is added or the existing hole is excited into a higher bonding orbital. A relaxed excited state of a defect can absorb photons by inducing electron transitions, hole transitions and the perturbed band-to-band transitions. Of these three, the latter two are most effective in weakening the bonds, the second one promotes holes to the inner bonding state and the third produces a new hole.

The Hattori model applies the Itoh-Nakayama mechanism to multi-hole localization at defect sites rather than perfect lattice sites. Localization of two-holes on surface defect sites can occur either by cascade excitation or by negative-U localization. Two-hole localization by these two processes differs from the *Knotek-Feibelman* (KF) two-hole localization induced by the e–hh Auger process initiated by core hole excitation [2.108]. The KF two-hole localized state is unstable against separation of the two holes because of the Coulomb repulsion, but in the negative-U two-hole localized state, the Coulomb repulsion energy is compensated by the lattice relaxation and instability may be on the lattice but not on the distance between two holes.

2.6.2 Calculational Techniques

The key advantage of focusing on the electronic processes which lead to laser ablation is that one gains the ability to compute the kinetics of ablation based on the fundamental physical mechanisms of desorption where one can hope to use physically measurable cross sections. In this section we present two examples of this concept.

One possible approach to the calculation of the microscopic dynamics of laser-induced desorption in weak-coupling solids is the bond–orbital theory of *Harrison* [2.109]. Recently, *Haglund* et al. applied this model to compute the fluence threshold for a number of compound semiconductors for which data on UV ablation threshold exist [2.110].

The fundamental idea of the bond–orbital theory is to construct matrix elements between the appropriate orbitals, then use experimental data to develop an appropriate set of universal constants which describe the systematic variations in matrix elements across a family of elements or compounds. The energy per bond pair in the perfect crystal can be computed in a number of ways.

For example, it may be derived from the moment distribution of the density of states using linear combinations of atomic orbitals to be [2.111]:

$$E_{\text{pair}} = -n\left(M_2 - \frac{M_4 - M_2^2}{4M_2}\right)^{1/2} + nV_0 + E_{\text{pro}}, \tag{2.2}$$

where n is twice the coordination number, M_2 and M_4 are, respectively, the second and fourth moments of the Electronic Density Of States (EDOS), V_0 is the repulsive screening potential and E_{pro} is the energy needed to create hybridized sp^3 bonds in the covalent solid. The second moment M_2 and the screening potential V_0 are both functions of the radial coordinate r. The fourth moment is a function of the tetrahedral bond angle, and hence depends on the details of surface bonding.

For laser-induced particle emission from GaP surfaces, the energy per bond pair at the surface is calculated in three-fold surface coordination. The computation of the energy per atom pair as a function of r in effect produces a cut through the radial potential energy surface. Parameters for GaP were optimized to reproduce the observed equilibrium interatomic spacing and bulk modulus for a four-fold coordinated solid. Bond-breaking in a three-fold coordinated surface site was then simulated by reducing the coordination number n by half the number of holes created in the bond.

Figure 2.13 shows how the radial dependence of the energy per pair for GaP, in a three-fold surface coordinated site, changes when first one hole and then two holes are generated in one of the bonds. While the surface-coordinated site is bound, and the one-hole state weakly bound, the two-hole state clearly represents an anti-bonding potential curve. Assuming that the creation of the holes is instantaneous, the excess positive energy

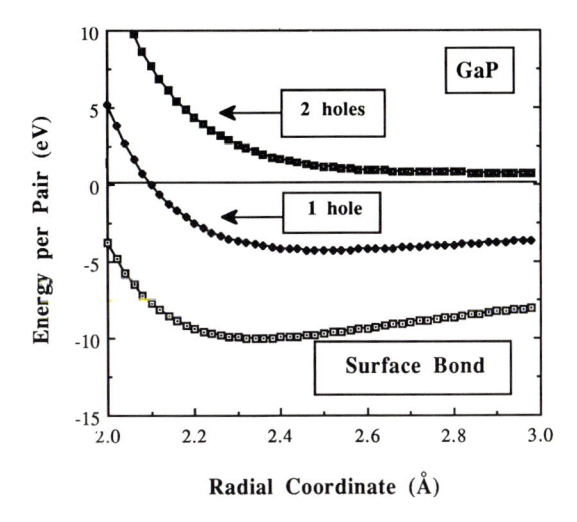

$$E_{\text{LR}} = E_{\text{pair}}^{2h}(r_0) - E_{\text{pair}}^{0h}(r_0) \tag{2.3}$$

Fig. 2.13. Dependence of the radial potential distribution for a three-fold coordinated surface site in GaP with zero, one and two holes in the bond. Note that the one-hole distribution is still weakly bound, while the two-hole potential curve is antibonding

at the equilibrium distance r_0 is available for lattice relaxation processes, including particle emission or sputtering. Implicit in this equation, of course, is the dependence of E_{LR} on the microscopic potentials calculated from the bond–orbital model. Thus E_{LR} is a function of the particular atomic site configuration at which the bond is broken.

However, the existence of a large relaxation energy is not of itself sufficient to guarantee that two-hole localization and/or particle emission will occur. Lattice relaxation must also provide greater energy gain for the system than delocalization in a band state, which means that we must have $E_{LR} > B/2$, where the bandwidth B is calculated from the second term in brackets in (2.1). Thus the conditions which determine the possibility of particle ejection from the lattice via the localization of two holes at a single lattice site are

$$E_{LR} < \frac{B}{2} \approx \frac{1}{2}\left(\frac{M_2 - M_4^2}{4M_2}\right)^{1/2} \text{ and } U_{eff} \equiv U_{sc} - E_{LR} < 0 , \qquad (2.4)$$

where the screened Coulomb potential U_{sc} is calculated as in the preceding section [see (2.1)].

The bond–orbital model also yields an estimate of the angular rigidity of a tetrahedral bond in a three-fold coordinated site as defined in [2.110] for Si, GaAs, GaP, GaN and ZnSe, all for the case of a single hole in a bond. Plotting this angular rigidity as a function of the effective potential U_{eff} gives a qualitative measure of the relative bond-breaking probability in these materials, since the creation of the second hole in a bond in any of them guarantees desorption. As shown in Fig. 2.14, the trends in angular rigidity parameter agree reasonably well with the overall trends for measured laser ablation thresholds for Group IV, III–V and II–VII materials. The bond–orbital approximation is not appropriate for strong-coupling solids unless one makes further adjustments to the density-of-states calculations [2.111].

Semi-empirical tight-binding-type quantum mechanical calculations, such as CNDO, are another approach for computing bond strengths and bond breaking energies near defects on semiconductor surfaces. *Khoo* et al. [2.107] used the

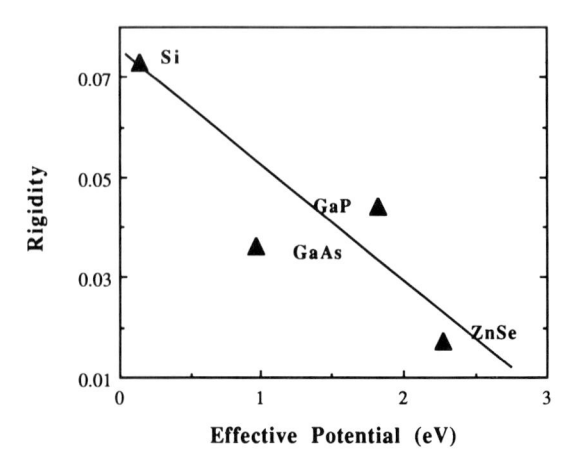

Fig. 2.14. Dependence of the angular rigidity parameter calculated from bond–orbital theory as a function of the effective potential energy U_{eff} for a number of elemental and compound semiconductors

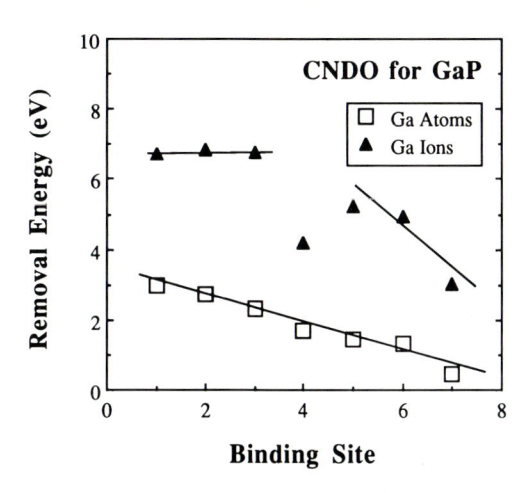

Fig. 2.15. Energies for removal of Ga atoms and ions from a GaP cluster, calculated in the CNDO approximation. The numbers on the abscissa correspond to: (1) Ga vacancy; (2) perfect lattice site; (3) step site; (4) Ga adatom; (5) P vacancy; (6) Ga–P divacancy; and (7) P(Ga) antisite defect. The smooth curves are simply guides for the eye

CNDO method to elucidate the effects of excitation of a Ga adatom on the GaP (1 1 0) surface and a Ga antisite defect situated in the second layer of the GaP (1 1 0) surface. They calculated the adiabatic potential energy surfaces for several excited states of these defects. For Ga adatoms, a second excitation leads to breaking the bonds for the adatom and consequently to particle emission, at an energy cost which varies according to the nature of the site (Fig. 2.15). An antisite defect is known to be a double acceptor. Hence, at the ground state, it includes two holes on the adatom which is localized on the antisite atom in the second layer. It is found that after relaxation which follows excitation of a hole, the Ga atom atop the antisite atom is displaced strongly (by 0.7 Å) outward from the surface; in this excited state the holes are localized more on the Ga atom on the surface. Although successive excitations cannot be calculated because of convergence problems, it is conceivable that further excitation finally leads to the bond–breaking.

The CNDO calculation indicates clearly that cascade excitation of defects on surfaces can cause rearrangement of holes or generation of new holes and eventually yields a catastrophic change in the atomic structure of the surface. Thus the multi-hole localization mechanism is likely to be the cause of defect-initiated desorption. Using CNDO, *Ong* et al. [2.112] also calculated the bond strength of Ga atoms at several defect sites (Fig. 2.15) and showed that higher laser fluence is needed to eject more strongly bound atoms.

2.7 Simulation of Laser Ablation

We have previously defined laser ablation to involve large-scale material removal (~ 1 ML/shot), threshold behavior and superlinear dependence of ablation yield on excitation density. Understanding laser-induced desorption is a prerequisite for dealing with laser ablation, both because surface conditioning at low fluence may affect the behavior in the ablation regime, and because the onset phase of a single laser pulse at above-threshold fluence may produce the same effects. Hence, laser-induced desorption may create the preconditions for laser ablation, but desorption and ablation need not proceed via the same mechanism.

Several phenomenological laser ablation models have been applied to specific materials under particular conditions. Most of them incorporate electronic excitation and relaxation processes only in an *ad hoc* fashion. Here, we present a defect-accumulation model of laser ablation into which the electronic effects of laser-induced particle emission and electronic structure can readily be incorporated. The defect-accumulation model takes into account the changes in the surface layers during a laser pulse due to the operation of electronic processes. Since the desorption mechanism – and hence the defect-accumulation phase – differs for solids with strong and weak electron–lattice coupling, the mechanisms of ablation or defect-accumulation for these two types of solids are also necessarily distinct and are treated accordingly.

2.7.1 Models of Laser Ablation

Most of the models proposed for laser ablation suffer in one degree or another from over-simplifications: unwarranted extrapolation from particle emission to ablation, failure to take sufficient account of the differences between bulk and surface conditions, or overlooking specific electronic effects which are material dependent. Nevertheless, many previous models have useful features to contribute to our general discussion. In this section, we review the principal features of several different models, and suggest ways in which the insights from each may be incorporated into the more general view of laser ablation. Our particular concern with each model will be the extent to which it provides a physical, material-dependent clue to the experimentally observed characteristics of ablation: rapid material removal, existence of a threshold and a nonlinear dependence of yield on fluence or laser intensity.

The oldest models for laser ablation were the thermal and electron-avalanche models. We mention them here primarily for completeness; their inadequacies have been documented in a number of papers [2.5, 7, 24]. The primary grounds for assuming thermal ablation are that nanosecond laser pulses are long compared to typical thermal diffusion times. It is therefore to be expected that if there is strong absorption and concomitant local heating, particles will boil off the surface at a rapid rate, leading to ablation. A thermal process can be assumed if the yield of ablated material is determined by the thermal yield integrated over the laser pulse:

$$Y \sim \sum_j \int dt\, m_j \alpha_j P_j (2\pi m_j k_B T)^{-1/2}. \tag{2.5}$$

Here the α_j are vaporization coefficients for the jth species, $P_j(\cdots)$ the vapor pressures, and T the absolute temperature [2.113]. While (2.5) follows in straightforward fashion from classical thermal physics, it seems to fit almost no experimental situation, except long-pulse laser ablation of metals and elemental semiconductors.

The electron-avalanche model, proposed and elaborated by *Bloembergen* and co-workers and summarized by *Smith* [2.114], assumes that destruction of the surface occurs by field-emission electron avalanches initiated by the high laser field at impurity inclusions and mechanical imperfections in surfaces or optical coatings. While the model successfully explained the problems with optical components which were failing because of poor-quality optical coatings, it failed to describe particle emission (desorption) or ablation from the high-quality, high-purity materials which have been increasingly studied in recent years. The avalanche model seems best suited to describe ablation due to extrinsic defects.

One of the earliest models of laser ablation based on electronic properties of materials was propounded by *Namiki* and co-workers [2.115] in order to explain the differences in laser-ablation thresholds among various semiconductors. The model is an intellectual descendant of the ideas of *Van Vechten* and others [2.49, 52, 53] in that it is based on the concept of dense electronic excitation due to the high probability of one-photon transitions in semiconductors. However, Namiki's idea stays closer to the idea of an electronic effect by assuming that there is a phase change or dimerization which occurs at high excitation density. The tetrahedral semiconductor bond is assumed to break when the bond angle is pushed beyond its elastic limit. In the Phillips bond-charge model, the transition from ionic to tetrahedral covalent bonding occurs for ionicity parameters of 0.78, so that the ionic II–VI semiconductors are predicted to have the lowest threshold, and the group IV semiconductors the highest threshold, in agreement with experiment. The relationship obtained by *Namiki* et al. can also be valid for the defect-initiated model described in Sect. 2.7.2.

Careful recent work on the mechanism of bulk laser damage in ultrapure materials for laser wavelengths such that $hv < E_{gap}$ has shown conclusively that avalanche breakdown is not the *intrinsic* damage mechanism in the bulk of most optical materials [2.116]. The mechanism of damage in this case is instead free-electron heating following multiphoton excitation [2.23, 24], with the damaged volume being heated beyond its yield strength by the temperature spike. Of great relevance to laser ablation are the observations by *Soileau* and co-workers [2.117] that suggest that self-focusing effects must be properly taken into account when dealing with *extrinsic* laser-induced damage – a category which should include laser ablation.

Reif, Matthias and their collaborators have proposed that defects accumulated during a laser pulse could enhance electron–hole pair generation by multiphoton *resonant* excitation and ionization in the alkaline-earth halides MgF_2, CaF_2 and BaF_2. In their early work, cluster calculations by *Reif* et al. [2.118] showed a large density of unoccupied surface electronic states which could be reached by multiphoton transitions. Measurements of electron and metal-ion yields showed a persistent resonance structure suggestive of excited-electronic states, and this, in turn, suggested a correlation between the intensity dependence of the yields and the number of photons required to reach these surface states [2.119]. The relatively defect-free surfaces of MgF_2 and CaF_2 have much lower multiphoton-excitation cross sections and, correspondingly, higher laser-ablation thresholds. This qualitative agreement should not obscure the fact that the calculation of the cluster surface states is not easy [2.120], so that detailed comparison of experiment and theory may still be premature.

In view of the concept that localization of electronic excitation is the crucial step for an efficient electronic-to-vibrational energy transfer, *Hattori* et al. [2.27] have suggested that emission of weakly bonded atoms in the topmost surface layer around vacancies can cause laser ablation. In this picture, the emission of these weakly bonded atoms causes the evolution of vacancy clusters in the topmost layer and this evolution progressively exposes the weakly bonded atoms originally in the second layer to the surface. Thus, the emission of weakly bonded atoms is assumed to expand the damaged layer both laterally and vertically.

A few comments on the validity of the model are in order here. First, it has been shown experimentally that repeated irradiation of the same spot on a GaP (1 1 0) surface at fluences below the ablation threshold – as determined by the modification of the LEED pattern – induces an increase in the yield, while it induces a decrease below the threshold, even if the emission is detected with a sensitivity of 10^{-4} monolayer [2.27]. This result cannot be explained if laser ablation is accompanied by a phase change, but it can be accounted for by a more microscopic alteration in the surface. According to the vacancy mechanism, the ablation threshold is that laser fluence at which the probability of ejecting weakly bonded atoms near vacancies approaches unity. Thus, the result described before does not favor the vacancy mechanism. Furthermore, although *Namiki* et al. have pointed out that the laser ablation is due to a phase change on the basis of the relation between the ionicity and the ablation threshold, as we pointed out already, the same relation should hold for breaking the bonds of weakly bonded atoms near the vacancies. Although further experimental proof is needed, we tentatively propose that the defect-initiated model holds for ablation of solids with weak electron–lattice coupling.

2.7.2 Model Calculations of Laser Ablation

The *résumé* of the previous section indicates that many models of laser ablation have been proposed for specific materials. In this section, we present a defect-accumulation model and show how it can be applied consistently to both strong- and weak-coupling solids.

In strong-coupling solids, such as alkali halides and alkaline-earth fluorides, laser-ablation studies are carried out with photons with energies smaller than the band gap. In these solids, multiphoton band-to-band transitions in the bulk produce Frenkel defects as well as electron–hole pairs and excitons. We suggest that emission of atoms by one-photon transitions related to the defects near the surface can enhance emission and consequently cause ablation. In solids with

2. Electronic Processes in Laser Ablation of Semiconductors and Insulators 43

weak electron–lattice coupling, this mechanism is not effective, since production of the Frenkel pairs is not induced by multiphoton band-to-band transitions. As shown in Sect. 2.6.1, desorption in these materials is defect related.

In NaCl and KCl, experiments show that electron–hole pairs or excitons generated by multiphoton absorption cause the emission of energetic halogen atoms when the electron–hole pairs or excitons are formed near the surface. Alkali atoms are evaporated thermally and, possibly, by direct photoexcitation of surface F-centers. The same photoexcitation process produces Frenkel pairs in the halogen sublattice when electron–hole pairs and excitons are generated in the bulk.

In the following model calculation, we assume that the stoichiometry in the surface layers is maintained under laser irradiation. We deal only with the rate-determining process, for example, the emission of a halogen atom. The secondary reactions under laser irradiation, such as emission of alkali atoms, is treated as subsidiary. This assumption should hold for model calculations, since the emission of alkali atoms from alkali-rich surfaces under laser irradiation is more feasible than emission of halogen atoms from the perfect surface. We accept the presumption that accumulation of defects in the surface layers induces resonant multiple-photon excitation to create F–H center pairs. In this case, since two defect pairs are generated close to each other, we assume that resonant multiple-photon excitation in deeper regions can cause emission of atoms.

Based on these arguments, we divided the alkali-halide crystal into three layers, as shown in Fig. 2.16. Region I consists of those few atomic layers from which decomposition of the self-trapped exciton leads to emission of a halogen and an alkali atom. Region II includes subsurface layers from which single-photon excitation due to the H-centers can cause emission. Region III is below Region II, and there F-centers and H-centers are simply accumulated. As the surface is eroded, Region II is converted to Region I, and Region III to Region II. The depth of Region I is governed by two factors: the range of surface layers

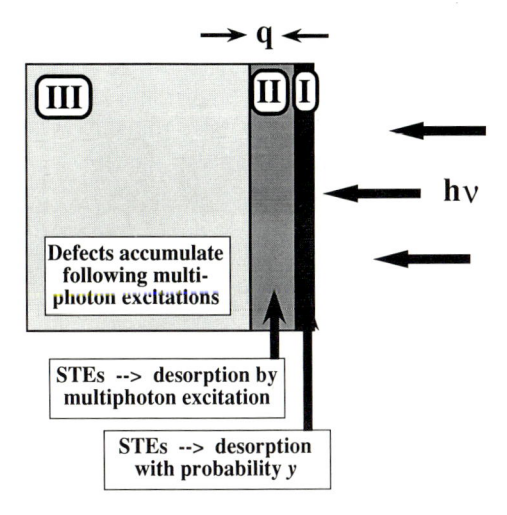

Fig. 2.16. Regions of a strong-coupling solid used in calculating the kinetics of laser ablation. As ablation progresses, the recession of the surface converts Region II into Region I, and Region III into Region II

within which the STE has an adiabatic instability, and by the diffusivity of excitons and ambipolar diffusion rate of electron-hole pairs. We take the thickness of Region I to be a monolayer. The thickness of Region II is denoted by q, which is kept constant with the erosion of the surface because the yields of F–H pairs in Regions II and III are the same.

Under these admittedly somewhat idealized conditions, the rate of emission dn_e/dt is given by

$$\frac{dn_e}{dt} = \frac{dn_S^I}{dt} + \sigma \Phi n_H \tag{2.6}$$

where the first term on the right-hand side is the rate of halogen-atom emission by multiple-photon excitation in Region I, n_H is the concentration of the H-centers per unit area in Region II, σ is the single-photon cross section for excitation of the H-center leading to photoinduced desorption and Φ is the laser flux density. Further,

$$\frac{dn_S^I}{dt} = \sigma^{(n)} \Phi^n N + \sigma_R^{(n)} \Phi^n n_H \tag{2.7}$$

where $\sigma^{(n)}$ is the n-photon absorption cross section, N is the number of host atoms per unit area and $\sigma_R^{(n)}$ is the resonant n-photon cross section of the F centers left after the emission of halogen atoms. The rate of generation of the H-centers in Region II is then given by

$$\frac{dn_H}{dt} = \frac{dn_S^I}{dt}(1 + q) \ . \tag{2.8}$$

This includes the contribution from Region II, and takes into account the conversion of Region III to Region II arising from erosion of the surface.

If we now introduce reduced time and concentration units $T = \sigma^{(n)} \Phi^n t$ and $n = n_I^j/N$, we find that the only free parameters are q and $R = \sigma_R^{(n)}/\sigma^{(n)}$. The resulting rate equations were solved using a Runge-Kutta method; the results are shown in Figure 2.17. Clearly, the desorption yield is proportional to Φ^n at the initial stage of irradiation, but it increases as Φ^{2n} at later times (i.e., at progressively larger total photon doses). When the resonant multiphoton excitation term is not included in (2.6), the yield scales as $\Phi^{1.5n}$.

This model is clearly too simple to explain all the processes which are involved in ablation. What is significant is that the characteristic features of the yield–fluence relation shown in Fig. 2.10 can be reproduced by a simple model. A more realistic level of complexity in the model can be obtained by incorporating defect-assisted multi-phonon transitions, interactions between defects on surfaces, formation of the laser-induced plasma, and so on.

We now consider what happens in weak-coupling solids, exemplified by the III–V compound semiconductor GaP following *Okano* et al. [2.121]. Since excitons in GaP are not self-trapped, desorption must arise entirely from defect

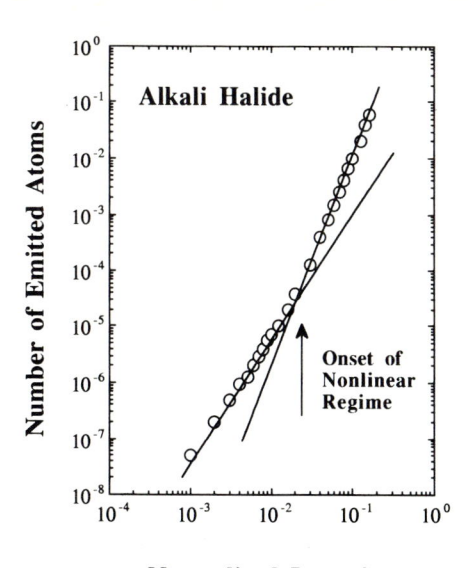

Fig. 2.17. Yield vs fluence relation for laser ablation from NaCl in a simple defect decay model, showing the existence of the linear and nonlinear regimes. The parameters for the calculation are found in Table 2.3

Table 2.3. Parameters for Laser-Ablation Calculation in a Strong-Coupling Solid

Physical parameter	Value
Two-photon absorption cross section $\sigma^{(2)}$	5.4×10^{-50} cm^2 s^{-1}
Resonant excitation cross section σ_{res}^2	4×10^{-46} cm^2 s^{-1}
Photodesorption cross section of H-center σ_h	10^{-19} cm^2
Thickness of Region II q [nm]	3
Laser-pulse width [ns]	10^{-8}
Order of multiphoton absorption process n	2

sites. Here, we assume that laser-induced desorption *below the ablation threshold* arises from defect sites only, and that laser ablation is induced by vacancy sites on the surface. The atoms surrounding the vacancy sites are more weakly bound to the lattice and thus most likely to be emitted during pulsed-laser irradiation, thus forming continuously evolving vacancy clusters. The growth of vacancy clusters on the surface layer increases the number of WBAs, thus enhancing particle emission. Furthermore, as vacancy clusters evolve in the top layer, vacancies in the second layer are either exposed or can be created, and the geometric multiplication of vacancy clusters in the second layer commences. Hence this model accounts for the rapid rise of ablation yield with fluence above threshold, and also for rapid lateral and vertical evolution of the damaged surface. Under these assumptions, the rate of emission of atoms is given by

$$\frac{dn_e}{dt} = \sigma^{(n)} \Phi^n n_D(t) , \qquad (2.9)$$

where Φ is the laser flux density, $n_D(t)$ is the number of WBAs around vacancies and vacancy clusters at time t, and n is the multiplicity required to break the bond for the WBA near vacancies. Note that n, in this case, does not necessarily refer to the number of photons necessary to produce the initial excitation; especially in the case of semiconductors and nanosecond pulses, the number n ranges from 2 to 6.

For small vacancy clusters, emission of one atom produces one WBA on average; thus

$$\frac{dn_e}{dt} = \frac{dn_D}{dt} \Rightarrow n_D(t) = n_D^0 \exp(\sigma^{(n)} \Phi^n t) \ . \tag{2.10}$$

As the vacancy cluster grows to be nearly circular, it can be shown that the increase in the number of WBAs is proportional to time (dose) [2.122]. Thus the time dependence of the number of WBAs is assumed to be

$$\frac{n_D(t)}{N_0} = z \exp(t) \ , \qquad \text{for } t < t_D \tag{2.11a}$$

$$= 2\pi(t - t_0) + z \exp(t_0), \text{ for } t > t_D \ , \tag{2.11b}$$

where N_0 is the number of vacancies per unit area initially present on the surface and z corresponds to the number of sites adjoining the vacancy cluster. We have also used the same reduced intensity units as above. A Monte Carlo calculation of this vacancy-induced ablation model for the surface top layer showed that the change from an exponential- to a power-law dependence occurs around $t_0 = 1.5$.

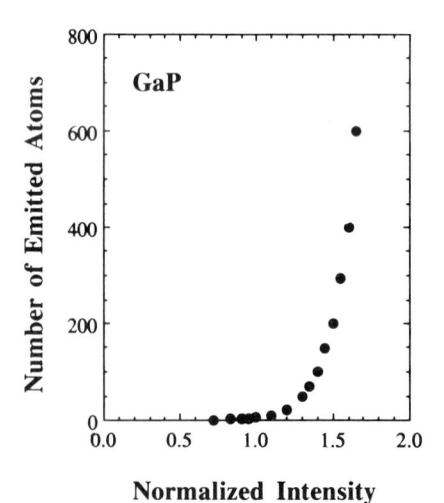

Fig. 2.18. Yield vs fluence relation for laser ablation from GaP calculated from a vacancy-cluster model

During a laser pulse of width τ_w, the number of atoms emitted from each vacancy can be found by simple integration of

$$\frac{n_e(\tau_w)}{N_0} = \sigma^{(n)} F^n \int_0^{\tau_w} \frac{n_D(t)}{N_\sigma} dt \ . \tag{2.12}$$

Figure 2.18 shows the relation between n_e/N_σ and $\tau^{1/n}$ for $n = 5$, indicating the yield versus fluence curve in reduced units. The sharp initial rise with yield on fluence can be reproduced by the present model of defect-initiated ablation with simple physical pictures of the elementary processes leading to ablation. A Monte Carlo simulation for a multilayer system gives nearly the same result. According to this model, the ablation threshold is given by $\sigma^{(n)} \Phi^n t_w \sim 1$, where the probability of breaking the bond for WBAs is approximately 1.

2.8 Summary and Conclusions

This survey of laser ablation has been concerned with the role of electronic processes in laser ablation from insulators and semiconductors. From our perspective, it is increasingly necessary and possible to start from the fundamental electronic excitation processes and follow the electronic and vibrational relaxational processes which produce surface decomposition and particle emission. This point of view allows one to begin true kinetic studies of ablation, and we expect such studies to assume increasing importance in the future.

We have described the fundamental mechanisms of laser ablation by a model which views laser ablation as the final, catastrophic outcome of the creation of vacancies by the four-step process which characterizes the precursor process of laser-induced desorption:

— initial electronic excitation creating electron–hole pairs,
— relaxation of the electron–hole pair to produce a localized excited state,
— evolution of the relaxed excited state to an intermediate excited configuration, and
— the transition by a non-radiative process to an anti-bonding potential energy surface.

We have shown that this *schema* makes it possible to identify the kinetic factors which allow the calculation of structural parameters (e.g., angular rigidity parameters) and rate constants associated with laser ablation.

The survey of experimental data indicates that experiments relating to the growth of vacancy and other defect clusters are particularly important. High-sensitivity measurements near the laser-ablation threshold are needed to clarify the roles of the electronic processes in precursor stages to massive ejection of surface atoms. Up to now, there are few measurements of the non-metallic species desorbed from insulators, even though this is widely believed to be the primary process. By using laser-induced fluorescence, it is already possible to follow the ejection of non-metallic species – halogens and oxygen, for example – and it is to be hoped that there will be more such measurements in the future. Resonance-ionization mass spectrometry has been shown to be capable of providing this information for many other species and its use will undoubtedly be expanded in future high-sensitivity measurements.

Characterization of the surface before and after desorption and ablation is another crucial requirement for understanding the transition from the desorption to the ablation regimes – a notable difficulty with insulators, but certainly possible at present for semiconductors by both traditional surface spectroscopies and scanning-tunneling microscopy. At this point, systematic measurements of weak-coupling solids can and should be made to test the models proposed here. With the rapid development of scanning microscopies such as atomic-force and photon-tunneling microscopies, it appears that the time is ripe to exploit ultrasensitive techniques for surface-structure measurements on insulators as well.

By classifying solids according to the strength of their electron–lattice coupling, it should also be possible to identify the most promising candidates as model systems for experiments. We note that oxide materials offer not only a wide range of possible electron–lattice couplings, but also, because of the range of bandgaps, permit the possibility of studying desorption and ablation at varying densities of electronic excitation. Of the potential model materials, the one which appears most undeservedly neglected up to now is SiO_2 in both its crystalline and amorphous forms. Experiments on other oxides under ultrahigh vacuum conditions with extreme sensitivity of detection should also help to clarify the role played by variations in electron–lattice coupling strength.

Acknowledgements. We express our gratitude to many colleagues for sharing data, insights and helpful criticisms with us. One of us (RFH) gratefully acknowledges support from the U.S.–Japan Cooperative Research Program of the National Science Foundation (INT-89-12697) and the Office of Naval Research under the Medical Free-Electron Laser Program, Contract N00014-91-C-109. Work at Nagoya University (NI) is partially supported by a grant-in-aid for specially promoted science of the Ministry of Education, and by the Japan Society for the Promotion of Science.

References

2.1 J.C. Miller, R. F. Haglund, Jr. (eds.): *Laser ablation – Mechanisms and Applications*, Lect. Notes Phys., Vol. 389 (Springer, Berlin, Heidelberg 1991)
M. von Allmen: *Laser-Beam Interactions with Materials*, Springer Ser. Mater. Sci., Vol. 2 (Springer, Berlin, Heidelberg 1987)
I.W. Boyd, E. Fogarassy, M. Stuke (eds.): Proc 1990 Spring Meeting of E-MRS, Appl. Surf. Sci. **46** (1990)
2.2 J.M. Hicks, L.E. Urbach, E.W. Plummer, H.-L. Dai: Phys. Rev. Lett. **61**, 2588 (1988)

2.3   W. Ho: Surface Photochemistry, in *Desorption Induced by Electronic Transitions DIET IV*, ed. by G. Betz, P. Varga, Springer Ser. Surf. Sci., Vol. 19 (Springer, Berlin, Heidelberg 1990) p. 48
2.4   F. Hillenkamp, B.T. Chait, R.C. Beavis, M. Karas: Anal. Chem. **63**, 1193A (1991)
       M.L. Gross (ed.): *Mass Spectrometry in the Biological Sciences: A Tutorial* (Kluwer, Dordrecht, 1991)
       A.L. Burlingame, J.A. McCloskey (eds.): *Biological Mass Spectrometry* (Elsevier, Amsterdam 1991)
2.5   E. Matthias, T.A. Green: Laser Induced Desorption, in *Desorption Induced by Electronic Transitions DIET IV*, ed. by G. Betz, P. Varga, Springer Ser. Surf. Sci., Vol. 19 (Springer, Berlin, Heidelberg 1990) pp. 112
2.6   R. Kelly, A. Miotello, B. Braren, A. Gupta, K. Casey: Nucl. Instrum. Methods B **65**, 187 (1992)
2.7   P. Kelly (ed.): *Laser Modification of Materials*, Opt. Eng. **28**, 1025 (1989)
2.8   N. Itoh: Nucl. Instrum. Methods Phys. Res. B **27**, 155 (1987)
2.9   The existence of the threshold for ablation or desorption is one of the most frequently remarked features of these processes. Because of the superliner character of the yield-fluence relation in ablation, it is not always clear whether the threshold is a manifestation of a critical process, a multiphoton excitation process, or simply of strong fluence dependence in the primary process. Nevertheless, the term fluence threshold is used for convenience throughout the paper, recognizing that use of the word does not, in itself, signify any particular physical mechanism
2.10   R.S. Knox: *Theory of Excitons* (Academic, New York 1963)
2.11   R.E. Howard, R. Smoluchowski: Phys. Rev. **116**, 314 (1959)
2.12   D. Pines: *Elementary Excitations in Solids* (Benjamin, New York 1964)
2.13   R.A. Smith: *Semiconductors* (Cambridge Univ. Press, Cambridge 1978)
2.14   C.H. Henry, D.V. Lang: Phys. Rev. B **15**, 989 (1977)
2.15   W. Hayes, A.M. Stoneham: *Defects and Defect Processes in Nonmetallic Solids* (J Wiley, New York 1985) Chap. 4
2.16   M. Ueta, H. Kanzaki, K. Kobayashi, Y. Toyozawa, E. Hanamura (eds.): *Excitonic Processes in Solids*, Springer Ser. Solid-State Sci., Vol. 60 (Springer, Berlin, Heidelberg 1986) Chap. 4
2.17   H.B. Nielsen, J. Reif, E. Matthias, E. Westin, A. Rosen: Multiphoton-Induced Desorption from BaF_2 (1 1 1), in *Desorption Induced by Electronic transitions DIET III*, ed. by R.H. Stulen, M.L. Knotek, Springer Ser. Surf. Sci., Vol. 19 (Springer, Berlin, Heidelberg 1986) p. 266
2.18   A. Okano, K. Hattori, Y. Nakai, N. Itoh: Surf. Sci. **258**, L671 (1991)
2.19   J. Kanasaki, A. Okano, K. Ishikawa, Y. Nakai, N. Itoh: Phys. Rev. Lett. **70**, 2495 (1993)
2.20   F. Manghi, C.M. Bertoni, C. Calandra, E. Molinari: Phys. Rev. B **24**, 6029 (1981)
2.21   A.M. Stoneham: *Theory of Defects in Solids* (Clarendon, Oxford 1975) Chap. 3.6
2.22   R.W. Dreyfus, F.A. McDonald, R.J. von Gutfeld: J. Vac. Sci. Technol. B **5**, 1521 (1987)
2.23   K. Hattori, A. Okano, Y. Nakai, N. Itoh, R.F. Haglund, Jr.: J. Phys. Condens. Matter **3**, 7001 (1991)
2.24   S.V. Garnov, A.S. Epifanov, S.M. Klimentov, A.A. Manenkov, A.M. Prokhorov: Izv. Akad. Nauk SSSR, Ser. Fiz. **52**, 1135 (1988)
2.25   S.C. Jones, P. Braunlich, R.T. Casper, X.-A. Shen, P. Kelly: Opt. Eng. **28**, 1039 (1989)
2.26   D. Arnold, E. Cartier, D.J. DiMaria: Phys. Rev. B **45**, 1477 (1992)
       D. Arnold, E. Cartier: Phys. Rev. B **46**, 15 102 (1992)
2.27   K. Hattori, A. Okano, Y. Nakai, N. Itoh: Phys. Rev. B **45**, 8424 (1992)
2.28   Y. Toyozawa: In *Vacuum Ultraviolet Radiation Physics*, ed. by E.E. Koch, R. Haensel, C. Kunz (Pergamon, Braunschweig 1984) p. 317
2.29   E.I. Rashba: *Excitons*, ed. by E.I. Rashba, M.D. Sturge (North-Holland, Amsterdam 1982) p. 543
2.30   Y. Toyozawa: J. Phys. Soc. Jpn. **50**, 1861 (1981)
2.31   P.W. Anderson: Phys. Rev. Lett. **34**, 953 (1975)
2.32   R.T. Williams, K.S. Song: J. Phys. Chem. Solids **51**, 679 (1990)
2.33   K. Kna'no, K. Tanaka, Y. Nakai: Rev. Solid State Sci. **4**, 383 (1990)

2.34 N. Itoh, K. Tanimura: J. Phys. Chem. Solids **51**, 717 (1990)
2.35 N. Itoh, K. Tanimura, A.M. Stoneham, A.H. Harker: J. Phys. Condens. Matter **1**, 3911 (1989)
2.36 J.M. Langer: Rev. Solid State Sci. **4**, 297 (1990)
2.37 T.N. Theis, P.M. Mooney, S.L. Wright: Phys. Rev. Lett. **60**, 361 (1988)
2.38 N. Itoh, T. Nakayama: Phys. Lett. A **92**, 471 (1982)
2.39 A.M. Stoneham: Rep. Prog. Phys. **44**, 79 (1981)
2.40 R. Englman: *Non-Radiative Decay of Ions and Molecules in Solids* (North-Holland, Amsterdam 1979)
2.41 W.P. Dumke: Phys. Lett. **78A**, 477 (1980)
2.42 E.J. Yoffa: Phys. Rev. B **21**, 2415 (1980)
2.43 N. Itoh: Adv. Phys. **31**, 491 (1982) and K.S. Song and R.T. Williams, *Self-Trapped Excitons* (Springer, Berlin, Heidelberg 1993)
2.44 A.M. Stoneham: Theory of Defects in Solids (Clarendon, Oxford 1975) Chap. 10.10
2.45 D.L. Dexter, C.C. Klick, G.A. Russell: Phys. Rev. **100**, 603 (1956)
2.46 T. Tokizaki, T. Makimura, H. Akiyama, A. Nakamura, K. Tanimura, N. Itoh: Phys. Rev. Lett. **67**, 2701 (1991)
2.47 L.C. Kimerling: Solid State Electronics **21**, 1391 (1978)
2.48 J. Schou: Nucl. Instrum. Methods Phys. Res. B **27**, 188 (1987)
2.49 J.A. van Vechten: In *Cohesive Properties of Semiconductors under Laser Irradiation*, ed. by L.D. Laude (Martinus Nijhoff, The Hague 1983) p. 49
2.50 D.J. Chadi, K.J. Chang: Phys. Rev. Lett. **60**, 2187 (1988)
      J. Dabrowski, M. Sheffler: Phys. Rev. B **40**, 10391 (1989)
2.51 A. Shluger, M. Georgiev, N. Itoh: Philos. Mag. B **63**, 955 (1991)
2.52 A. Van Vechten, R. Tsu, F.W. Saris: Phys. Lett. **74A**, 422 (1979)
2.53 M. Wautclet, L.D. Laude: Appl. Phys. Lett. **36**, 197 (1980)
2.54 D. Pooley: Solid State Commun. **39**, 241 (1965)
2.55 R.T. Williams, K.S. Song, W.L. Faust, C.H. Leung: Phys. Rev. B **33**, 7232 (1986)
2.56 K. Tanimura, T. Suzuki, N. Itoh: Phys. Rev. Lett. **68**, 635 (1992)
2.57 D.L. Bouche-Bruevich, E.G. Landsberg: Phys. Status Solidi (a) **29**, 9 (1968)
2.58 W.L. Brown: In *Laser and Electron-Beam Processing of Materials*, ed. by C.W. White, P.S. Peercy (Academic, New York 1980) pp. 20
2.59 H. Kanzaki, T. Mori: Phys. Rev. B **29**, 3573 (1984)
      N.D. Stoffel, R. Riedel, E. Colavita, G. Margaritondo, R.F. Haglund, Jr., E. Taglauer, N.H. Tolk: Phys. Rev. B **32**, 6805 (1985)
2.60 A. Schmid, P. Braunlich, P.K. Rol: Phys. Rev. Lett. **35**, 1382 (1975)
2.61 T.A. Green, G.M. Loubriel, P.M. Richards, N.H. Tolk, R.F. Haglund, Jr.: Phys. Rev. B **35**, 781 (1987)
      M. Szymonski, P. Czuba, T. Dohnalik, L. Jozefowski, A. Karawajczyk, J. Kolodziej, R. Lesniak: Nucl. Instrum. Methods B **48**, 534 (1990)
2.62 M. Szymonski, J. Kolodziej, P. Czuba, P. Piatkowski, A. Poradzisz, N.H. Tolk, J. Fine: Phys. Rev. Lett. **67**, 1906 (1991)
2.63 X. Li, R.D. Beck, R.L. Whetten: Phys. Rev. Lett. **68**, 3420 (1992)
2.64 A. Schmid: University of Rochester (unpublished data)
2.65 C.R. Phipps, Jr., T.P. Turner, R.F. Harrison, G.W. York, W.Z. Osborne, G.K. Anderson, X.F. Corlis, L.C. Haynes, H.S. Steele, K.C. Spicochi: J. Appl. Phys. **64**, 1083 (1989)
2.66 J. Reif: Opt. Eng. **28**, 1122 (1989)
2.67 J. Valbis, N. Itoh: Rad. Eff. Def. Solids **116**, 171 (1991)
2.68 R.W. Dreyfus, R. Kelly, R.E. Walkup: Appl. Phys. Lett. **49**, 1478 (1986)
2.69 R.W. Dreyfus, F.A. McDonald, R.J. von Gutfeld: Appl. Phys. Lett. **50**, 1491 (1987)
      R.J. von Gutfeld, F.A. McDonald, R.W. Dreyfus: Appl. Phys. Lett. **49**, 1059 (1986)
2.70 J.E. Rothenburg, R. Kelly: Nucl. Instrum. Methods B **1**, 291 (1984)
2.71 L.L. Chase: Laser ablation and optical surface damage, in *Laser Ablation: Mechanisms and Applications*, ed. by R.F. Haglund, Jr., J.C. Miller, Lect. Notes Phys., Vol. 389 (Springer, Berlin, Heidelberg 1991) pp. 193

2.72 M.A. Schildbach, A.V. Hamza: Phys. Rev. B **45**, 6197 (1992)
2.73 D.L. Griscom: Nature of defects and defect generation in optical glasses, in *Radiation Effects in Optical Materials*, SPIE Proc. **541**, 38 (1985)
C. Itoh, K. Tanimura, N. Itoh, M. Georgiev: Rev. Solid State Sci. **4**, 679 (1990)
2.74 T.T. Tsai, D.L. Griscom, E.J. Friebele: Phys. Rev. Lett. **61**, 444 (1988)
2.75 R.A.B. Devine: Phys. Rev. Lett. **62**, 340 (1989)
2.76 M. Rothschild, D.J. Ehrlich, D.C. Shaver: Appl. Phys. Lett. **55**, 1276 (1989)
2.77 J.T. Dickinson, S.C. Langford, L.C. Jensen, P.A. Eschbach, L.R. Pederson, D.R. Baer: J. Appl. Phys. **68**, 1831 (1990)
2.78 Z.A. Rachko, J.A. Valbis: Phys. Status Solidi (b) **93**, 161 (1979)
2.79 R.L. Webb, L.C. Jensen, S.C. Langford, J.T. Dickinson: J. Appl. Phys. **74**, 2323 (1993)
2.80 R.E. Leuchtner, K.S. Grabowski, D.B. Chrisey, J.S. Horwitz: Appl. Phys. Lett. **60**, 1193 (1992)
2.81 R.F. Haglund, Jr., J.H. Arps, K. Tang, A. Niehof, W. Heiland: MRS Proc. **201**, 507 (1992)
2.82 R.F. Haglund, Jr., M. Affatigato, K. Tang, C.H. Chen: Appl. Phys. Lett. (to be published)
R.F. Haglund, Jr., K. Tang, C.H. Chen: MRS Proc. **244**, 405 (1992)
2.83 N.S. Nogar, R.C. Dye, R.C. Estler, S.R. Foltyn, R.E. Muenchausen, X.D. Wu: Diagnostic studies of YBa$_2$Cu$_3$O$_{7-\delta}$ laser ablation, in *Laser Ablation: Mechanisms and Applications*, ed. by R.F. Haglund, Jr., J.C. Miller, Lect. Notes Phys., Vol. 389 (Springer, Berlin, Heidelberg 1991) pp. 3 and references therein
2.84 T. Nakayama, N. Itoh, T. Kawai, K. Hashimoto, T. Sakata: Rad. Effects Letters **67**, 129 (1982)
2.85 Y. Nakai, K. Hattori, A. Okano, N. Itoh, R.F. Haglund, Jr.: Nucl. Instrum. Methods B **58**, 452 (1991)
2.86 K. Ichige, Y. Matsumoto, A. Namiki: Nucl. Instrum. Methods B **33**, 830 (1988)
2.87 J.C. Phillips: *Bonds and Bands in Semiconductors* (Academic, New York 1973)
2.88 A. Okano, J. Kanasaki, Y. Nakai, N. Itoh: J. Phys. Condens. Matter, in press (1994)
2.89 Y. Kumazaki, Y. Nakai, N. Itoh: Phys. Rev. Lett. **59**, 2883 (1987)
Y. Kumazaki, Y. Nakai, N. Itoh: Surf. Sci. Lett. **184**, L445 (1987)
2.90 J.M. Moison, M. Bensoussan: J. Vac. Sci. Technol. **21**, 315 (1982)
2.91 T. Nakayama: Surf. Sci. **133**, 101 (1983)
2.92 J. Kanasaki, A. Okano, K. Ishikawa, Y. Nakai, N. Itoh: J. Phys. Condens. Matter **5**, 6497 (1993)
2.93 A. Okano, K. Hattori, Y. Nakai, N. Itoh: Surf. Sci. Lett. **258**, L671 (1991)
2.94 N. Itoh, A. Okano, K. Hattori, J. Kanasaki, Y. Nakai: Nucl. Instrum. Methods B **82**, 310 (1993)
2.95 J.P. Long, S.S. Goldenberg, M.N. Kabler: Phys. Rev. Lett. **68**, 1014 (1992)
2.96 P.D. Brewer, J.J. Zinck, G.L. Olson: Appl. Phys. Lett. **57**, 2526 (1990)
2.97 D. Menzel, R. Gomer: J. Chem. Phys. **41**, 3311 (1964)
P.A. Redhead: Cdn. J. Phys. **42**, 886 (1964)
2.98 N. Itoh, Y. Nakai, K. Hattori, A. Okano, J. Kanasaki: In *Desorption Induced by Electronic Transitions-DIET V* ed by A.R. Burns, E.B. Stechel, D.R. Jennison, Springer Ser. Surf. Sci., Vol. 31 (Springer, Berlin, Heidelberg 1993) pp. 233
2.99 P.D. Townsend, F. Lama: In *Desorption Induced by Electronic Transitions-DIET I*, ed. by N.H. Tolk, M.M. Traum, J.C. Tully, T.E. Madey, Springer Ser. Chem. Phys., Vol. 24 (Springer, Berlin, Heidelberg 1983) pp. 220
2.100 M. Szymonski, J. Kolodziej, P. Czuba, P. Piatkowski, A. Poradisz, Z. Postawa: Nucl. Instrum. Methods B **65**, 507 (1992)
2.101 V. Puchin, A. Shluger, N. Itoh: Phys. Rev. B **47**, 10 760 (1993)
2.102 P.H. Bunton, R.F. Haglund, Jr., D. Liu, N.H. Tolk: Phys. Rev. B **45**, 4566 (1992)
E. Taglauer, N. Tolk, G. Margaritondo, N. Gershenfeld, N. Stoffel, J.A. Kelber, G. Loubriel, A.S. Bommanavar, M. Bakshi, Z. Huric: Surf. Sci. **169**, 267 (1986)
2.103 N. Itoh, K. Tanimura, C. Itoh: Self-Trapped Excitons in Amorphous and Crystalline SiO$_2$, in *The Physics and Technology of Amorphous SiO$_2$*, ed. by R.A.B. Devine (Plenum, New York 1988) pp. 135
2.104 N. Itoh, T. Nakayama, T.A. Tombrello: Phys. Lett. A **108**, 480 (1985)

2.105 H. Sumi: Surf. Sci. **248**, 382 (1991)
2.106 Z. Wu: Laser-Induced Ion Emission from Si and Ge Surfaces, In *Desorption Induced by Electronic Transitions DIET IV*, ed. by G. Betz, P. Varga, Springer Ser. Surf. Sci., Vol. 19 (Springer, Berlin, Heidelberg 1990)
2.107 G.S. Khoo, C.K. Ong, N. Itoh: Phys. Rev. B **47**, 2031 (1993)
2.108 M.L. Knotek, P.J. Feibelman: Phys. Rev. Lett. **40**, 964 (1978)
2.109 W.A. Harrison: *Electronic Structure and the Properties of Solids* (Dover, New York 1989)
2.110 R.F. Haglund, Jr., K. Hattori, Y. Nakai, N. Itoh: Proc. of the Workshop on Beam–Surface Interactions, OSA Technical Digest Series, Vol. 10 (1991)
2.111 W.A. Harrison: Phys. Rev. B **41**, 6008 (1990)
2.112 C.K. Ong, G.S. Khoo, K. Hattori, Y. Nakai, N. Itoh: Surf. Sci. Lett. **259**, L787 (1991)
2.113 R.W. Dreyfus, R. Kelly, R.E. Walkup: Appl. Phys. Lett. **49**, 1478 (1986)
2.114 N. Bloembergen: IEEE J. QE **10**, 375 (1974)
W.L. Smith: Opt. Eng. **17**, 489 (1978)
2.115 A. Namiki, S. Cho, K. Ichige: Jpn. J. Appl. Phys. **26**, 39 (1987)
2.116 X.A. Shen, P. Braunlich, S.C. Jones, P. Kelly: Phys. Rev. B **38**, 3494 (1988)
2.117 M.J. Soileau, W.E. Williams, N. Mansour, E.W. Van Stryland: Opt. Eng. **28**, 1133 (1989)
2.118 J. Reif, H.B. Nielsen, O. Semmler, E. Matthias, E. Westin, A. Rosén: J. Vac. Sci. Technol. B **5**, 1415 (1987)
2.119 J. Reif, H. Fallgren, W.E. Cooke, E. Matthias: Appl. Phys. Lett **49**, 770 (1986)
J. Reif, H. Fallgren, H.B. Nielsen, E. Matthias: Appl. Phys. Lett. **49**, 930 (1986)
2.120 E. Westin, A. Rosén, E. Matthias: Molecular Cluster Calculations of the Electronic Structure of the (1 1 1) Surface of CaF_2, in *Desorption Induced by Electronic Transitions DIET IV*, ed. by G. Betz, P. Varga, Springer Ser. Surf. Sci., Vol. 19 (Springer, Berlin, Heidelberg 1990)
2.121 A. Okano, A.Y. Matsuura, K. Hattori, N. Itoh, J. Singh: J. Appl. Phys. **73**, 3158 (1993)

3. Laser Ablation and Optical Surface Damage

L.L. Chase

With 17 Figures

Optical surface damage is a very complex and important problem in the development and application of high-power lasers. Optical damage may also be considered to be the initial phase of laser ablation. The fundamental mechanisms of optical damage are therefore of significance for applications of laser ablation; on the other hand, measurements of laser ablation have provided valuable evidence regarding the fundamental mechanisms of optical damage. Other methods for investigating optical surface damage include surface characterization before and after the damage event, short-pulse optical probes of laser–surface interactions, photothermal deflection or deformation, and transient grating techniques to measure heating and thermal conduction of surfaces and coatings. These methods are discussed and the current state of understanding of optical damage mechanisms is reviewed.

3.1 Introductory Remarks

Optical surface damage adversely affects the performance, cost, and reliability of lasers and optical systems more than any other factor. Although this problem has existed since the invention of the laser, the causes of optical damage remain uncertain in many cases, despite three decades of research. There are two main difficulties involved in investigating the causes of optical damage: i) damage occurs above a threshold, below which it is difficult to find evidence of any precursor phenomena, and ii) other events that are initiated by the primary damage mechanism often obscure the nature of the initial laser–surface interaction. These are illustrated in Fig. 3.1, which shows an interference (Nomarski) micrograph of the optical damage sites on BK-7 silicate glass irradiated at a fluence just above its damage threshold by a 1 ns long laser pulse at a wavelength of 1.06 μm. Small pits appear, surrounded by shallow ablated regions, which are highlighted in contrast by the interference technique. This shallow ablation is caused by a plasma formed from the materials ejected from the pit. The initiation process is therefore the one which caused the pit, not the one which produces most of the ablation. For many materials, the surface damage morphology is not so easily interpreted, and it is not clear what features were caused by the initial damage event. Similar uncertainties are involved in interpreting the results of experiments to investigate damage mechanisms.

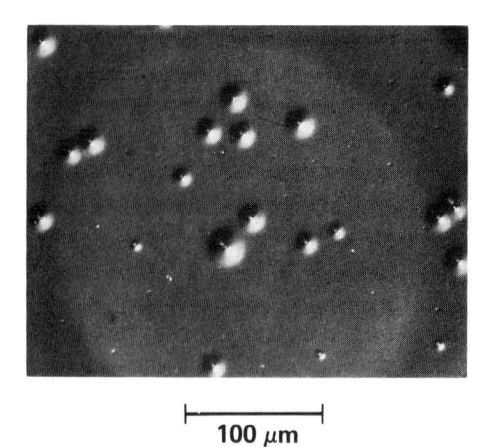

Fig. 3.1. Optical damage morphology obtained with Nomarsky interference microscopy on the exit surface of a BK-7 borosilicate glass sample irradiated at a fluence of 25 J/cm² by a 1 ns pulse at a wavelength of 1054 nm. The nominal damage threshold is 17 J/cm²

100 μm

The challenge involved in understanding the causes of optical damage and surface ablation is to identify the relevant interactions of the laser beam with the surface and to investigate how these interactions are related to surface properties. This challenge is difficult to meet with straightforward measurement methods because of the threshold nature of optical damage. Some progress has been made recently with three strategies. The first approach is to search for evidence of physical interactions, such as desorption processes, occurring below the damage threshold and to connect them to the optical damage event in some way (for example, by their spatial coincidence with a subsequent optical damage event). A second method is to correlate optical damage thresholds with surface properties obtained using the conventional UHV spectroscopies and the recent scanned-probe techniques. The third approach involves laser pump–probe techniques that investigate the excitations produced on the surface by laser light. After a review of the characteristics and potential mechanisms for optical damage, the major topic of this article will be a discussion of these three approaches to investigating optical surface damage mechanisms.

It is impossible within the limited length of this article to provide an exhaustive review of all of the relevant material or to cite all of the literature of the preceding three decades of optical damage research. Unfortunately, there is no really exhaustive review of the subject, but the reader is referred to the book by *Wood* [3.1] and to the series of proceedings of the Boulder Conference on Materials for High Power Lasers (Boulder Damage Conference) which were initially published by the National Bureau of Standards (now the National Institute of Standards and Technology) as a special publication series. Some review articles appeared in a special issue of the journal Optical Engineering [3.2] and other book chapters [3.3, 4].

3.2 Characteristics of Optical Surface Damage

Optical surface damage is not a very well-defined phenomenon. Operationally, it is usually defined as a laser-induced surface feature that is observable in an optical microscope, because that implies that the optical performance of the surface is affected. The method of performing measurements of optical surface damage affects the measured value of the threshold. Care must be exercised to obtain reproducible optical pulses that are smoothly varying in space and time, and to accurately measure the maximum intensity of the laser pulse. This requires single-mode laser beams and good laser-beam diagnostics. Even if these measurement conditions are achieved, the measured damage threshold fluences can be influenced by: 1) laser pulse width and shape, 2) the size of the illuminated area, 3) random variations from site to site, and, 4) the number of laser pulses. The effects of these experimental conditions on the damage threshold provide very useful evidence regarding surface-damage mechanisms, in addition to affecting the performance of the material in various laser applications. It is worthwhile to discuss several of these characteristics of optical damage in more detail.

The dependence of optical damage thresholds on laser pulse duration is a very important and revealing effect. Optical damage thresholds are specified as either the fluence, or, alternatively, the intensity, at which damage is first observed. Neither method is "correct" because damage thresholds vary nonlinearly with laser pulse width. This is demonstrated for a set of optical glasses in Fig. 3.2, which shows that the damage threshold fluences F_d vary as the laser pulse width t_p, as $F_d \sim t_p^{0.4}$ [3.5]. Typically, this type of behavior has been observed for a variety of bare surfaces and coatings for pulse durations ranging from 0.1 ns up to and beyond the microsecond regime. The exponent of t_p varies for various materials from 0.2–0.7, with an average of about 0.35 or so [3.6]. If the peak intensity of the laser pulse determined the threshold fluence, it would be linear in the pulse duration, and if the fluence were the determining factor the threshold fluence would be independent of pulse duration. The observed dependence implies that neither the peak intensity nor the fluence of the pulse determines the damage threshold. Damage fluences that vary sublinearly with the pulse duration, as in Fig. 3.2, could result from a simple process like thermal diffusion, which can determine the maximum temperature reached in an absorbing region of the material, or perhaps from more complex processes that modify the surface characteristics during the laser pulse.

Figure 3.2 also illustrates another interesting characteristic of optical damage. The data in this plot were obtained at different times by different investigators on carefully ground, polished and cleaned glasses that exhibit a variety of chemical and physical characteristics. Nonetheless, they have very similar surface damage thresholds to within a factor of 50%, or so. This behavior is typical of a much larger set of wide bandgap insulating optical materials (with a somewhat larger range of variability of the damage threshold) and it is roughly

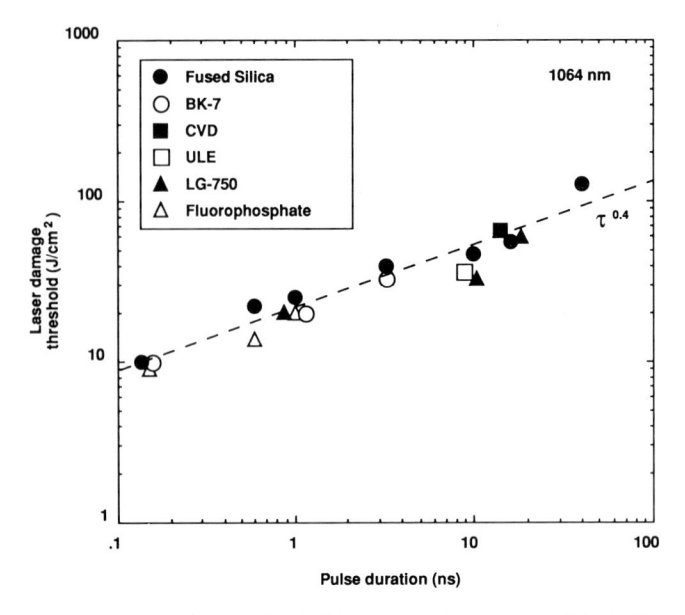

Fig. 3.2. Optical damage thresholds of a set of optically polished silicate, phosphate, and fluorophosphate glasses as a function of laser pulse duration at a wavelength of 1064 nm [3.5]

summarized by the rubric "the optical damage threshold is about 10 J/cm^2 at a pulse width of 1 ns at a wavelength of 1 μm". The materials' property that seems to correlate with the optical damage threshold in this class of materials is the refractive index: the higher the index, the lower the damage threshold. The refractive index is related to other properties, such as the optical bandgap energy and the enhancement of the local optical electric field around defects, so this correlation is not unambiguously interpretable, as will be obvious from the discussion to follow.

The area and/or the location of the irradiated region can influence the damage threshold if the damage is caused by defects, and the dimensions of the irradiated spot become smaller than the spacing between the defects. This behavior can be very pronounced for bulk damage; in quartz, for example, thresholds observed with well-focused laser beams are often orders of magnitude larger than those observed for mm or cm size regions [3.7]. On well-prepared surfaces, this effect is usually less extreme, although there are sizable variations of damage thresholds from site to site [3.8]. For this reason it is necessary to specify the measurement criteria when quoting damage thresholds. For example, the threshold might be defined as the fluence at which a certain percentage (usually 50% or less) of the irradiated sites are damaged. Clearly, a statistically adequate number of sites must be irradiated at a given fluence to determine such a threshold.

The effects of repetitive laser pulses on optical damage thresholds are very informative, and occasionally beneficial. Most often, however, the damage threshold decreases, sometimes by as much as an order of magnitude, with repetitive irradiation. This cumulative behavior of multiple pulses implies that during the "non-damaging" pulses some modification of the surface occurs which often cannot be detected in the microscope used for observation of damage. (Occasionally, however, small damage spots have been observed during repetitive irradiation of a region, and they differ from the usual optical damage behavior in that they do not "worsen" on succeeding pulses.) It is possible that this accumulation effect is related to the dependence of the damage threshold on pulse duration in some cases, since the surface may also be modified during a pulse. It has become customary in recent years to determine damage thresholds by multiple-pulse irradiation, since the multiple-pulse threshold is of greater importance than the single-pulse threshold for laser applications. In some cases, the damage threshold can be increased by irradiation with laser pulses below the damage threshold. A recent example of this "conditioning" effect is multilayer SiO_2–HfO_2 multilayer coatings, for which the damage threshold can be raised by as much as a factor of 2.5 by repetitive irradiation with pulses of gradually increasing fluence [3.9]. The probable causes of this effect, which involve microscopic surface modification at fluences below the single pulse damage threshold, will be discussed in a later section.

The morphology of optically damaged surface regions is dependent on the type of material and the laser pulse characteristics (wavelength, pulse duration, irradiated area, etc.). The regular-shaped, widely spaced damage spots in Fig. 3.1 are fairly typical of well-polished glass surfaces at wavelengths ~ 1 µm. For crystals, optical coatings, and for glasses irradiated at shorter wavelengths, there is a wide variation in the morphology, ranging from localized pits to widespread ablation over a large fraction of the irradiated area. Cratering, cracking, cleaving, exfoliation, melting, and spallation may occur, particularly at wavelengths much longer than the bandgap absorption. At UV wavelengths, a common morphology resembles that of etching of regions encompassing a large fraction of the irradiated area. This is possibly indicative of "electronic" or plasma-dominated ablation processes. The reader is referred to an interesting study of ablation morphology of sapphire at wavelengths ranging from 1.06 µm to 0.266 µm, which illustrates most of these morphologies [3.10, 11]. Very often, a hot plasma, visible as a spark on the surface, is generated at fluences above the damage threshold. It was once believed that such a plasma is required for optical damage to transparent materials, but counterexamples were found. Plasma phenomena during laser excitation and optical damage were studied photographically, and it was found that there were often two plasmas, one generated in the atmosphere and one generated in the materials ablated from the surface [3.12]. The plasma in the atmosphere was shown to be unrelated to the initiation of optical damage, since damage thresholds are essentially independent of the ambient atmosphere, including the case of a high-vacuum environment [3.13]. Although the morphology of the damaged regions may provide some useful information regarding the cause of the damage event, this information is often incomplete and ambiguous in nature.

3.3 Possible Causes of Optical Damage

There are a variety of possible physical mechanisms and surface characteristics that could be involved in optical surface damage. The materials used in high-power lasers and optical systems are usually chosen so that the laser photon energy is much smaller than the optical bandgap and much larger than optical phonon energies, so intrinsic linear absorption is negligible. There are, however, several other absorption mechanisms that can initiate optical damage, particularly on surfaces. Figure 3.3 summarizes these mechanisms for a typical optical material. The only intrinsic processes that can occur are MultiPhoton Absorption (MPA) and multiphonon absorption. At wavelengths shorter than a few micrometers, multiphonon absorption is too weak to cause sufficient energy deposition for optical damage through direct heating so we can neglect the multiphonon absorption. The 10.6 μm wavelength of the CO_2 laser, however, is very strongly absorbed by some materials and may cause optical damage.

Multiphoton absorption generates mobile carriers through interband transitions. Multiphoton absorption alone cannot deposit sufficient energy to cause optical damage directly, since the cross sections for these transitions are extremely small. For example, the largest Two-Photon Absorption (TPA) coefficients β are on the order of 10 cm/GW, and the rate at which energy is deposited per unit volume is given by βI^2. Since the highest surface damage thresholds for nanosecond pulses are about 10 J/cm^2 (I = 10 GW/cm^2), the energy deposition due to TPA is about 10 J/cm^3, sufficient to heat the material by a few degrees centigrade, at best. MPA is of greatest importance for optical damage processes because of the free carriers and absorbing centers that result from the photoexcited states that are produced.

Free carriers can be generated by MPA and by absorption induced by defects and impurities. These carriers can, in turn, absorb additional energy from the laser beam. If the free carriers absorb enough energy to heat the medium sufficiently to cause melting, vaporization, or fracture, optical damage can result. Alternatively, if the energy is absorbed at a rate higher than it can be transferred to the lattice via collisions, the carriers can gain enough energy to

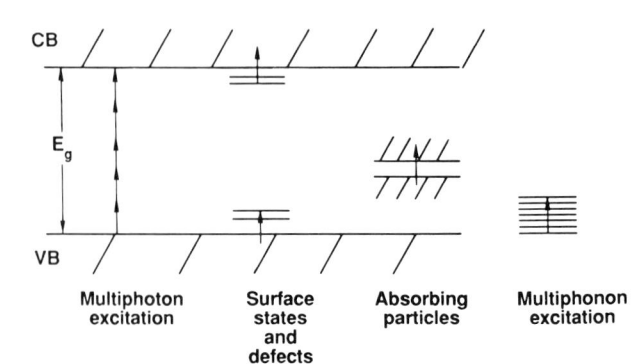

Fig. 3.3. Valence and conduction bands of a typical insulating optical material showing the intrinsic and extrinsic laser–surface interaction mechanisms

excite additional carriers across the bandgap, and an "avalanche" of ionization occurs, resulting in a carrier density that grows exponentially with time. This "optical breakdown" is analogous to dc or microwave breakdown of media, and the intrinsic optical breakdown strength establishes an upper limit for optical damage thresholds.

Absorption of light by free carriers requires the change in direction of the carrier velocity due to collisions; otherwise, the carriers simply oscillate in the optical field with no net increase in their kinetic energy. The absorption rate therefore depends on the ratio of the collision rate to the frequency of the electromagnetic field. For the simplest model of a single collision rate that is independent of carrier kinetic energy, the absorption rate varies as [3.14]

$$\frac{dE}{dt} = \frac{ne^2\tau_c E^2}{m(1 + \tau_c^2\omega^2)}, \tag{3.1}$$

where n is the carrier density, τ_c is the collision rate due to phonons, m is the effective mass, E is the optical electric field, and ω is the angular frequency. This shows that free carrier absorption increases with wavelength. In reality, the collision rate is energy dependent, so (3.1) is at best a rough approximation [3.14]. In a solid, for example, optical phonon scattering dominates at low carrier energies, but acoustic phonon scattering takes over and greatly increases the scattering rate before the avalanche threshold is reached (i.e., carrier energies larger than the bandgap) [3.15]. Monte-Carlo calculations for electron heating in fused silica show that the polar optical phonon scattering dominates for carrier energies up to about 3 eV, above which the acoustic phonon scattering rate rapidly becomes larger [3.15]. If carrier diffusion in energy space is considered, it has been argued that the free carrier absorption increases with laser intensity as $I^{3/2}$ [3.16–18]. This has been corroborated to some extent by pump–probe experiments in the bulk of fused silica and some alkali halides [3.19].

Since electron avalanche mechanisms have been repeatedly invoked as an optical damage mechanism, it is worth reviewing the long and checkered history of the understanding of this phenomenon. The spark that accompanies both surface and bulk optical damage of many materials shows that optical damage is often accompanied by avalanche breakdown. This may, however, be a natural consequence of the melting, vaporization, fracturing, and ionization that may accompany thermally initiated mechanisms of optical damage. The question to be addressed is whether avalanche ionization, starting from an initially negligible carrier density, can by itself cause sufficient energy deposition to initiate optical damage. Initial theoretical work on this topic was hindered by a lack of solid information or modeling capability regarding electron–phonon scattering rates for hot electrons, which vary widely with energy, temperature, and material properties [3.14]. Recent experimental and theoretical work has shed some light on this topic, although it is the author's opinion that this recent work is still not completely conclusive or comprehensive [3.19].

The recent theoretical work on the effects of an optical electric field on hot-electron dynamics in bulk fused silica by *Arnold* et al. is particularly enlightening [3.15]. They performed Monte-Carlo calculations of the energy changes as a function of time of a single electron in an intense optical electric field in fused silica. Using algorithms and parameters derived from studies of dc electric breakdown of silica, they calculated electron–phonon scattering rates, energy-absorption rates, avalanche ionization rates, and electron-energy distributions for hot carriers in fused silica. There calculations are reproduced in Figs. 3.4 and 3.5. Figure 3.4 shows that free carrier heating at intensities less than 1 TW/cm^2 (10^{12} W/cm^2) increases rapidly with temperature. On the other hand, the calculated avalanche ionization rate shown in Fig. 3.5 decreases dramatically with increasing temperature. More recent work on fused silica using a quantum mechanical treatment of the electron dynamics shows that the avalanche field is not so strongly temperature dependent as predicted by the classical calculations in Fig. 3.5 [3.20]. For materials with smaller electron–phonon coupling than fused silica, it is possible, however, that this temperature-dependent avalanche rate will occur, and free-carrier heating can suppress avalanche ionization. For this reason it has been argued that free-carrier heating alone can precipitate optical damage by melting or vaporizing the medium, while suppressing an avalanche in the carrier density [3.19]. This argument is supported by experiments on KBr and other alkali halides, which show that damage occurs because the inferred temperature reaches the melting point before an avalanche of ionization occurs [3.19]. Although the accuracy of these data might be questioned because of the indirect method used to measure the temperature,

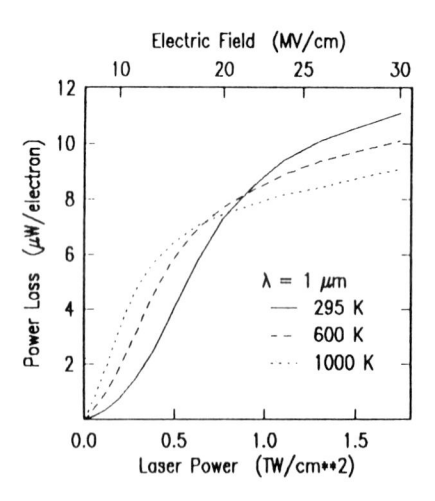

Fig. 3.4. Calculated power loss per conduction electron to the lattice in fused silica irradiated by a laser at a wavelength of 1 µm as a function of laser power for several lattice temperatures [3.20]

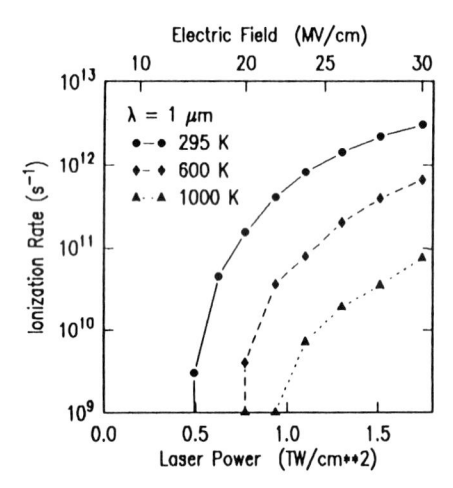

Fig. 3.5. Calculated impact ionization rate caused by a 1 µm laser as a function of laser power for several lattice temperatures [3.20]

they show that the role of avalanche ionization in bulk optical damage is subject to doubt.

The possible role of avalanche ionization in optical damage to surfaces is even more uncertain. The laser intensities ($< 10\,\text{GW/cm}^2$) reached near the surface-damage thresholds of most surfaces irradiated by nanosecond or longer laser pulses are too low to lead to electron avalanche in the bulk. The possibility exists, however, for ionization and breakdown in the region outside, but near to, the surface. Breakdown thresholds of gases can be far lower than those of solids. One can envision a chain of events in which ablated electrons, ions, and neutrals are further ionized and heated by the laser light via the inverse Bremsstrahlung process, and the resulting plasma bombards the surface to cause further ablation and heating. This feedback mechanism, which could ultimately culminate in optical breakdown, is also a potential cause of cumulative damage, since the bombardment of the surface can generate defects that increase the absorption of laser light on successive laser pulses.

The extrinsic absorption processes diagrammed in Fig. 3.3 are potentially very important for optical damage. Impurities or surface states can leave filled (donor) electronic states near the conduction band or empty states (acceptors) near the valence band. In both cases free carriers can be generated with one photon absorption if the binding energies are less than the photon energy. These carriers can function as the "seed" electrons for avalanche ionization or they can cause thermally induced damage if they are present in sufficient concentration for free-carrier heating to cause temperatures that exceed the melting point or temperature gradients that cause fracture.

The other type of defect depicted in Fig. 3.3 represents absorbing regions (often referred to as inclusions) that may be present near the surface due to foreign material left by processing, handling, or other contamination. These absorbing inclusions might also consist of non-stoichiometric regions in the material, caused by surface degradation or coating processes. Bulk damage thresholds in some optical materials are often correlated with the presence of absorbing particles, such as the platinum inclusions in laser glass melted in platinum crucibles.

There have been several calculations of the heating of absorbing inclusions by laser excitation [3.22–24]. The important characteristics of an inclusion are its size, absorption coefficient, thermal properties, and thermal contact with the medium. Large, highly absorbing inclusions of high thermal conductivity absorb light in a thin surface layer and rapidly transfer the deposited heat away into their bulk and into the surrounding matrix. Very small inclusions, on the other hand, also have a low ratio of heated volume to surface area and rapidly distribute the absorbed energy to the medium. These small inclusions also do not create the large thermal stresses that cause optical damage. It turns out that in dielectric media particles with dimensions ranging from tenths of the micron to about ten microns are capable of reaching the highest temperatures or causing the largest thermal stress at a given laser fluence. Other factors, such as the detailed thermomechanical properties and other physical effects of a heated inclusion on the medium, also influence optical damage thresholds. For example, it has been suggested that the blackbody radiation of a heated inclusion may be effective in producing electron hole pairs in the surrounding medium, thus precipitating breakdown or free-carrier heating [3.25]. Clearly, this topic is too complex for general conclusions to be drawn. It is important to realize, however, that optical damage may be caused by relatively weakly absorbing inclusions with dimensions too small for them to be observed in an optical microscope. This will be illustrated by an example in a subsequent section.

There is another class of defects that reduce the damage thresholds of surfaces and some bulk solids. These are dielectric discontinuities that enhance the local optical electric field. Examples of such defects have been given by Bloembergen, who calculated the electric field enhancements due to idealized models of cracks, voids, and surface pits [3.26, 27]. Table 3.1 gives the electric field enhancements of these defects in terms of the refractive index of the medium. The enhancements, which can reduce the damage threshold by up to an order of magnitude, increase rapidly with increasing refractive index. This is one possible reason why optical damage thresholds of glasses generally decrease with increasing refractive index. Defects as small as 100 nm may be effective in reducing damage threshold. Surfaces are particularly susceptible to these types of structural defects. Polished surfaces may have polishing scratches, pits, surface and subsurface cracks, and debris left behind from polishing and atmospheric exposure. Even vacuum-cleaved surfaces have such defects in the form of cleavage debris, steps, and cleavage cracks. In fact, it has been found that the surface damage thresholds of some well-polished surfaces are higher than those of the same material when it is cleaved. This would appear to provide strong evidence for the importance of local field enhancement mechanisms. Careful studies of the effects of surface roughness on the optical damage thresholds of glasses have also verified these effects for surface roughnesses down to 100 nm [3.28].

Optical coatings are particularly susceptible to optical damage involving both absorbing inclusions and dielectric discontinuities. Coatings are more likely than bulk materials to have nonstoichiometric regions that can be absorbing. They also are likely to have high densities of defects caused by columnar growth, grains, islands, and voids. It is therefore not surprising that optical damage to coated surfaces is the most pressing problem in high-power laser optics. The method and conditions of coating deposition are extremely important for achieving high damage thresholds.

Table 3.1. Electric field enhancement factors for three idealized dielectric discontinuties

Spherical void	$3n^2/(2n^2 + 1)$
Cylindrical groove	$2n^2/(n^2 + 1)$
Crack (disk shaped ellipsoid)	n^2

3.4 Investigation of Optical Surface Damage Mechanisms

In the previous section, a variety of possible causes and influencing factors that may contribute to optical surface damage were reviewed. It is likely that different types of materials are affected by different combinations of these factors, so the problem of understanding and controlling optical damage mechanisms is therefore likely to depend on the material, and even on different conditions (wavelength, pulse duration, etc.) for a given material. In this section, some of the promising methods for the investigation of optical damage mechanisms are discussed, with an emphasis on recent work.

3.4.1 Laser Ablation as a Probe of Optical Damage

Investigations of laser ablation have provided important insights into optical surface damage mechanisms. For many materials, desorption, of constituents of the medium, as well as impurities, occurs at fluences below the optical damage threshold [3.3]. This desorption can provide information about the laser–surface interaction mechanisms. Desorption is also a form of microscopic optical damage that may be related to the "visible" damage that occurs at higher fluences or on subsequent laser pulses. Laser-induced emission of neutral atoms and molecules, ions, and electrons has been observed from a variety of materials. In this review some of the results that have a bearing on optical damage mechanisms are briefly discussed.

The earliest work on laser ablation was reported by *Rousseau* et al. who observed both positively and negatively charged emission from optical materials following irradiation with a ruby laser [3.29]. They attributed this to heating and desorption of contaminants adsorbed onto the sample in their relatively low vacuum (10^{-5} Torr). *Schmid* et al. [3.30] were the first to demonstrate the utility of Laser-Induced Desorption (LID) for investigating laser–surface interactions. They observed the directional emission of halogen atoms from alkali-halide crystals following multiphoton excitation (up to four photons) by a ruby laser. The directionality of the emission result from the recoil of the halogen atom along (1 1 0) lattice directions following the nonradiative decay of a color center created by the absorption event. It was argued that this ablation could modify the near-surface region because of the loss of halogen ions and create defect centers that cause additional absorption of laser light, eventually leading to the optical damage which was observed after repetitive illumination of the same surface region [3.30].

There have been many subsequent investigations of laser-induced neutral and charged emission from nonmetallic crystals and glasses. The observed neutral particle ablation processes appear to fall into three classes. At the lowest fluences, desorption is strongest for the first laser pulse and often decreases rapidly on subsequent pulses at the same fluence. Typically, a slight increase in fluence will cause the emission to resume. In cases where velocity distributions have been measured, the effective temperatures of the emitted neutrals can range from the ambient surface temperature up to well above the vaporization temperature of the medium as the fluence increases. Figure 3.6 shows an example of a time-of-flight distribution for Na emitted from freshly air-cleaved NaF irradiated by a 5 ns, 266 nm laser pulse in this low fluence regime [3.31]. The effective temperature of this distribution is about 300 K, the same as the surface temperature. Data like those in Fig. 3.6 were also taken at 1064 nm fundamental and the 532 nm second-harmonic wavelengths of the Nd:YAG laser. The threshold fluence for this emission was only about a factor of two higher and the emission yields were comparable. These properties, together with the obvious threshold character of the emission show clearly that it is not caused by multiphoton absorption. Excitation of surface states or impurities is the most likely mechanism. It has been suggested that the desorption in this regime detaches a fixed number of loosely bound atoms (near ledges on cleaved surfaces, for example) or atoms residing near absorbing defects, which are "bleached" by the first few laser shots [3.31]. There are no reports of multiple-pulse optical damage occurring in this fluence regime, so it is not clear if these desorption events are related to optical damage.

In the second fluence regime, above the multiple pulse damage threshold, but often well below the single pulse threshold, the neutral emission yield rises rapidly, usually in a threshold-like manner, and surprisingly high particle energies are often observed, despite the absence of any hot plasma (see the inset of Fig. 3.6) [3.31]. Ions and electrons are also often emitted at these fluences. Figure 3.7 shows the yield of Na atoms emitted from air-cleaved NaF versus fluence. For NaF surfaces, the high particle energies have been correlated with

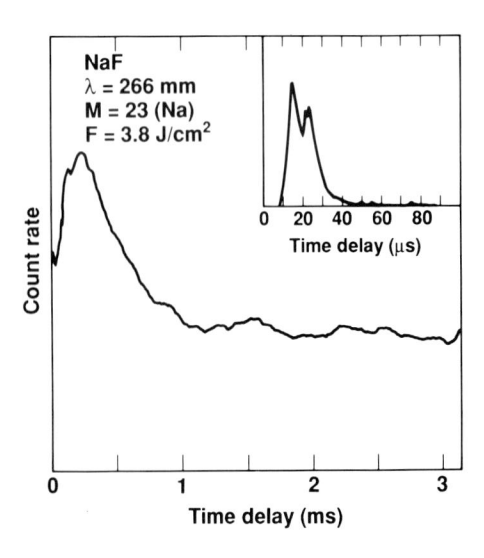

Fig. 3.6. Time-of-flight distribution for Na neutrals desorbed from an NaF cleaved surface by a 266 nm, 5 ns laser pulse at a fluence of 3.8 J/cm^2. The damage threshold is about 7.5 J/cm^2. Inset is a high energy component of the ablated Na neutrals and ions [3.31]

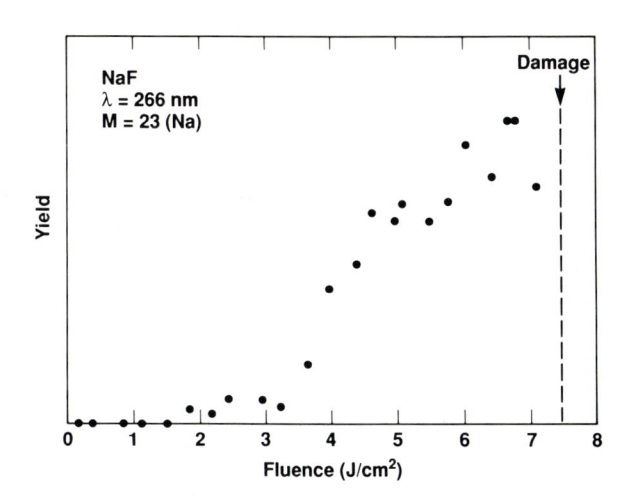

Fig. 3.7. Yield of emitted Na neutrals as a function of laser fluence for a cleaved NaF surface irradiated by 266 nm, 5 ns laser pulses [3.31]

microscopic cleavage of the sample [3.31]. It is likely that the energetic particles emitted from other materials in this fluence range are also caused by cleavage or fracture. The mechanism for this "fractoemission" is as yet uncertain, but it has been suggested to arise from acceleration in the potential of the large surface charges on fresh cleavage faces [3.32]. In this regime of fluences, one is clearly observing evidence of the laser–surface interaction that causes optical damage and/or cumulative surface modification leading to damage.

Finally, above the threshold for observable damage, there is usually a tremendous increase in the emission yield, a hot plasma spark is usually observed, and particle energies can exceed 100 eV, even for neutrals. These high-energy neutrals are believed to be produced by acceleration of ions in the space-charge electric field caused by the rapid expansion of the higher velocity electron density relative to the slower moving ions, and the ions are subsequently neutralized as they catch up with and pass through the electron cloud. Evidence for emission of high energy neutrals in this fluence regime due to cleaving and cracking has also been found [3.31]. Although this fluence regime is a very productive one from the point of view of laser ablation rate, it offers little opportunity to learn about the initial causes of optical damage. The sudden increase in the yield of emitted particles can, however, be used to detect the presence of optical damage quickly and sensitively.

Laser-induced emission of neutrals and ions below the optical damage threshold could occur through several mechanisms. Thermal vaporization is one possibility, which can be easily investigated since effective particle temperatures can be obtained from easily measured kinetic energy distributions. The particle temperature and its relation to laser fluence and thermal properties of the medium provides strong evidence for or against a thermal mechanism for the emission. For materials that are nominally transparent at the laser wavelength, thermally induced particle emission could result from heating caused by large clusters of absorbing defects, such as non-stoichiometric regions. If electronic excitations are produced by multiphonon absorption or absorption caused by surface states or defects, it is possible for surface atoms or ions to be desorbed if their binding potentials are reduced by the excitation. This type of process is referred to as "electronic" desorption, analogous to the desorption caused by core-level excitations as observed in the often-quoted work of *Knotek* and *Feibelman* on oxide crystals [3.33, 34]. In contrast with the case of core-level excitations, however, the evidence for the specific laser-induced electronic desorption mechanism involving valence electrons is very seldom conclusive.

Itoh and coworkers [3.35, 36] have suggested that, in materials where excitons are self-trapped, the recoil energy produced by their decay can cause desorption. The data of *Schmid* and *Braunlich* seems to be a clear example of this [3.30]. In materials where excitons are not self trapped it is argued that more than one hole is required on a surface anion to form a repulsive potential that will result in desorption [3.35, 36]. The localization of two holes at a single ionic site requires that the Coulomb repulsion be overcome. It has been suggested that either lattice relaxation or screening by a large excited carrier density can provide this effective "negative-U potential" [3.35, 36], but to the author's knowledge, no convincing experimental evidence or theoretical substantiation exists for this hypothesis. The need for a large screening carrier density is consistent, however, with the observation that a threshold fluence or intensity is required for desorption from such materials (two such examples will be discussed later). Quantitative theoretical and experimental work is needed in this area.

Zinc sulphide is an interesting material for the investigation of laser-surface interaction mechanisms. It has a band gap of 3.8 eV, so the fundamental and first three harmonics of a Nd:YAG laser span the range of photon energies from well below to well above the bandgap. It was observed that the threshold fluence for laser-induced neutral emission decreases rapidly with increasing photon energy, but it remains about a factor of three lower than the single-shot optical damage threshold over this range of wavelengths [3.31]. As the laser fluence is increased, a gradual heating of the kinetic energy distribution of the emitted neutrals occurs, and emission of very energetic particles, possibly due to fracture, is observed just below the single-pulse damage threshold. Experiments were also performed with excimer-laser irradiation at a photon energy above the bandgap, and with velocity distributions determined by Doppler broadening of a two-photon absorption line of atomic Zn [3.37]. The measured yield of neutral Zn as a function of particle temperature inferred from the Doppler measurements is shown in Fig. 3.8. The total yield of emitted Zn as a function of measured particle temperature was found to be consistent with sublimation of the ZnS due to surface heating, since the enthalpy ΔH is close to the enthalpy of sublimation. The experiments done on ZnS with excitation both above and below the energy gap have been interpreted as indicating that desorption of Zn neutrals from this material is thermally induced [3.31, 37]. This conclusion is supported by the results of laser pump-probe measurements, discussed later.

Another interesting series of laser-induced desorption experiments was done on the indirect gap crystal GaP using a tunable dye laser. It was found that

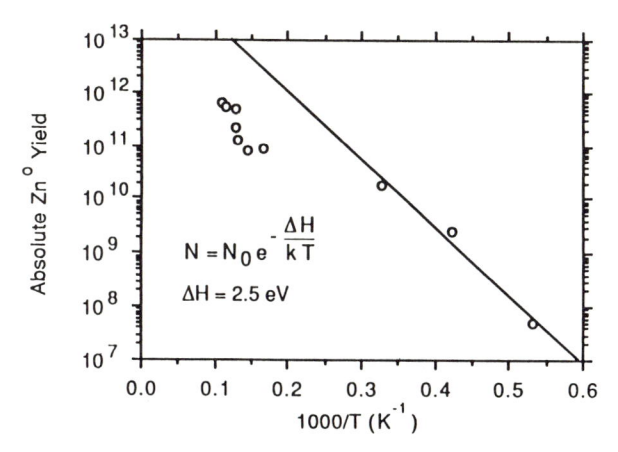

Fig. 3.8. Yield of Zn neutrals desorbed from a polished ZnS surface by 308 nm laser pulses. The temperature scale was obtained from measurements of Doppler broadening of a two–photon absorption line of Zn. The solid line is an Arrhenius plot with $\Delta H = 2.5$ eV, comparable with the dissociation enthalpy of ZnS [3.37]

excitation of surface states just below the bandgap enhances the desorption yield dramatically [3.38, 39]. It is interesting that, despite the apparent "electronic" nature of the desorption process including the involvement of a specific surface state, the desorption is observable only at fluences above a threshold. As was discussed before, the threshold-like behavior may be a consequence of the need for screening effects that require multiple electronic excitations in order for desorption to occur.

The laser-induced emission of charged species from optical materials has also been the subject of several investigations. *Siekhaus* et al. [3.40] studied the emission of electrons from a variety of media following nanosecond laser excitation below the damage threshold fluence under UHV conditions. They showed that the total emitted charge increased superlinearly with the laser fluence, as shown for several materials in Fig. 3.9. The slope of these log–log plots is roughly equal to the ratio of the work function of the materials to the photon energy, and it was suggested that the emission might be a multiphoton photoelectric phenomenon, which had been observed previously for metals. The total charge emitted from the insulators increased roughly as the square root of the pulse duration for a fixed laser intensity, however, and the total charge was so large that the surfaces would charge up to potentials approaching 100 keV at the highest laser intensities. In subsequent investigations, *Chase* and *Smith* [3.31] found that the emission consisted of equal magnitudes of positive ions and electrons. Large fluxes of neutrals were also observed from several of the media using a quadrupole mass spectrometer. It was concluded that the neutrals were emitted initially from the surface, followed by multiphoton ionization in the gas phase.

Some very interesting studies of charged emission were performed by *Matthias* and coworkers on vacuum-cleaved surfaces of CaF_2 and BaF_2 irradiated by a tunable dye laser at wavelengths in the range 410–450 nm [3.41–43]. In these mass- and charge-selected experiments, the positive ions F^+, Ba^+, and BaF^+ were emitted from BaF_2. The yield varied strongly with laser fluence and

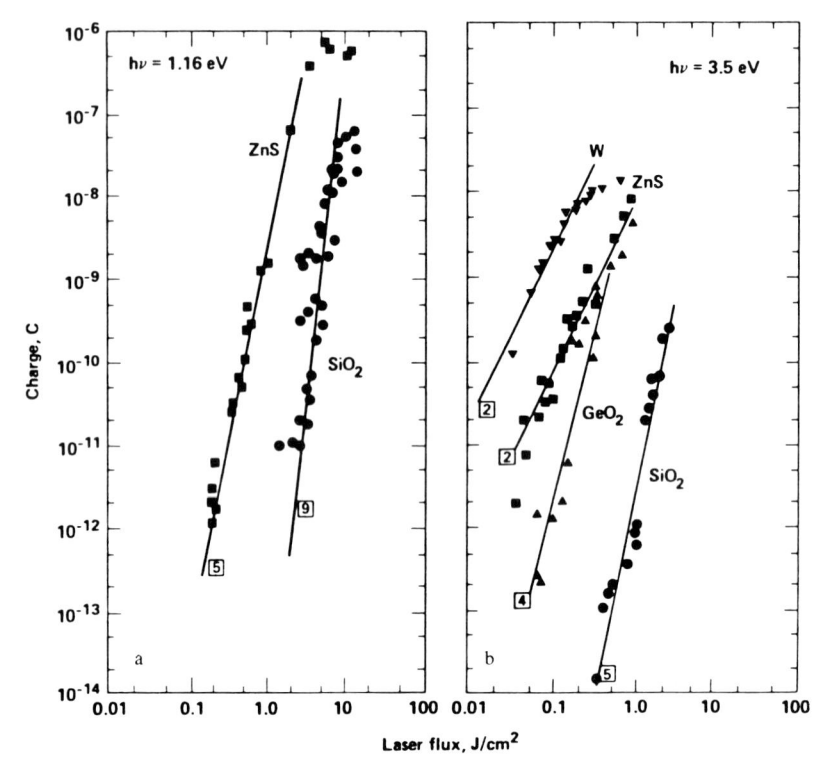

Fig. 3.9. Total charge emitted from the surfaces of several optical materials and W metal as a function of laser fluence for two laser energies. The solid lines are fits to a power law dependence [3.40]

wavelength, suggestive of multiphoton absorption involving real intermediate states. The authors found that the emission of Ba$^+$ and F$^+$ were anticorrelated and interpreted this as resulting from removal of successive Ba and F layers from the surface. It was suggested that the apparent involvement of five-photon absorption in these experiments could be due to a large enhancement of the usually very weak cross section by resonant intermediate states [3.44]. In contrast with the copious emission observed from vacuum cleaved surfaces of these fluoride crystals, it was reported that no charged or neutral desorption occurs from polished surfaces of these crystals [3.31]. This illustrates the sensitivity of laser desorption to surface conditions.

Sapphire (Al$_2$O$_3$) is a wide-bandgap optical material with very favorable surface properties for studies of surface morphology, surface electronic structure, and laser ablation processes. It has a high surface conductivity which minimizes charging, and it has a well-characterized surface morphology, with reconstruction of the (1 1 2 0) surface into a 12 × 4 structure occurring above about 1500 K. The average surface absorption of laser light by sapphire was investigated by *Dreyfus* et al. [3.45] using photothermal deflection, and it was found that detectable surface heating was observable only at wavelengths shorter than 350 nm, or so. *Schildbach* and *Hamza* studied laser ablation by nanosecond laser pulses at wavelengths of 1064 and 355 nm [3.46]. They observed the emission of Al$^+$ ions at fluences somewhat below the single-pulse damage threshold. The emission threshold, and the optical damage threshold are somewhat lower at 355 nm than at 1064 nm. This emission is also of the first type mentioned before, since it bleaches after a number of laser shots. The velocity distribution of the Al$^+$ ions, shown in Fig. 3.10, is very revealing. The solid curve is a fit to the observed time-of-flight data using a shifted Maxwellian distribution given by

$$f(t) = A_0 t^{-5} \exp\left[-A_1\left(\frac{s}{t} - A_2\right)^2\right], \qquad (3.2)$$

where t is the time, s is the flight distance, and A_1 is a measure of the width of the distribution. A_2 is a "stream velocity" that amounts to a center-of-mass kinetic

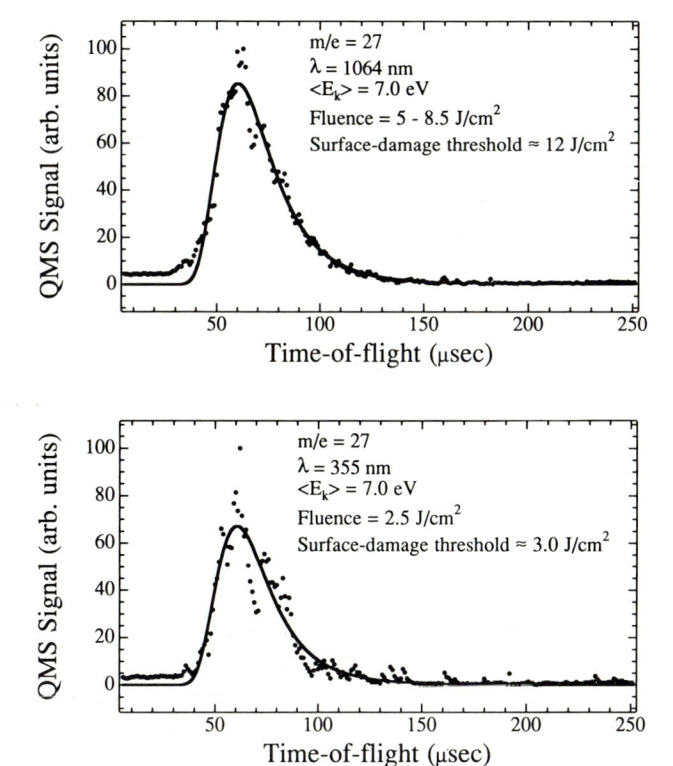

Fig. 3.10. Time-of-flight distributions of Al$^+$ ions desorbed from a 12 × 4 reconstructed sapphire surface by the fundamental and third harmonic of a Nd:YAG laser. Laser wavelengths are 1064 nm (*upper*) and 355 nm (*lower*). Solid curves are fits to (3.2) with an average particle energy of 7–8 eV [3.46]

energy of the ion plume of about 7.8 eV, which is reproducible, independent of laser fluence and laser wavelength. *Schildbach* and *Hamza* have interpreted this desorption as resulting from the excitation of an exciton that is localized on the Al ions [3.46]. This surface excitation reduces the binding potential of the Al ions and they are desorbed, taking with them as kinetic energy a large fraction of the electronic excitation energy of the surface exciton. This presumably accounts for the large, reproducible kinetic energy of the Al^+ ions. A major question is how the irradiation of a crystal with a bandgap of over 9 eV by photons with energy of 1.17 eV (1064 nm) results in the desorption of ions with energies of 7.8 eV. Some related surface analysis experiments on sapphire and a possible mechanism for the Al ions emission will be discussed in the next section.

A final illustration of the use of laser ablation measurements to study optical damage mechanisms is provided by some recent work on multilayer, high-reflectivity optical coatings. The coatings investigated are made of about thirty alternating layers of HfO_2 and fused SiO_2. It was found recently that the optical damage threshold of these coatings can be increased by up to a factor of over 2.5 by irradiating them with multiple pulses of laser light at fluences below the single-shot damage threshold [3.9]. This "conditioning" effect is essentially permanent, and it is a good example of one of the rare beneficial effects of cumulative modification of materials by laser pulses. *Schildbach* et al. [3.47] showed that when this conditioning occurs, large, random bursts of neutral Hf, Si and O are emitted. Sometimes these bursts persist for a number of shots, but they disappear if the fluence is below the single-pulse damage threshold. The laser threshold fluence required for the appearance of the ablation was shown to be close to the threshold fluence required for the laser conditioning. A typical time-of-flight distribution for the emitted neutral Hf on a single laser shot above the conditioning fluence is shown in Fig. 3.11. The shape of this distribution is very irregular, and it is centered at a time delay indicative of a very high particle kinetic energy, about 20 eV. Since there is no hot plasma observed during laser

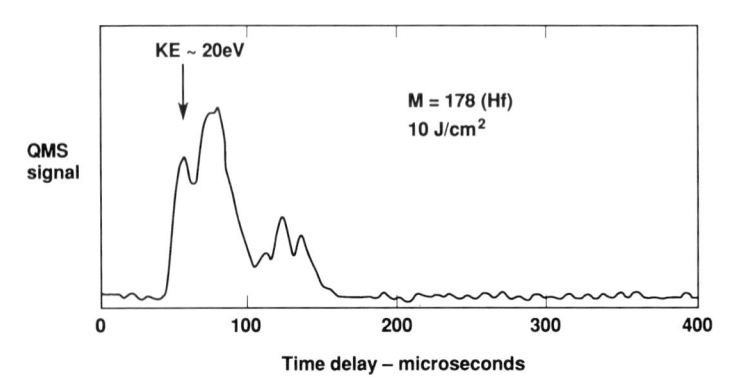

Fig. 3.11. Time-of-flight distribution of the neutral Hf atoms ablated from a multilayer HfO_2–SiO_2 optical coating illuminated by a 1064 nm laser pulse at a fluence of 10 J/cm^2 during "laser conditioning". The single–pulse damage threshold is about 16 J/cm^2

conditioning, it seems likely that these high energies are again the result of fractoemission due to cracking of the coating during conditioning. It was suggested that the conditioning effect results from this fracture because of either the ejection of optically absorbing materials from the coating or from stress relief due to the cracking. Laser ionization mass analysis of the coatings was also done, and it was concluded that there were not impurities in the optically damaged regions of the sample. This suggests that physical defects or non-stoichiometry cause the optical damage. Further evidence of the conditioning mechanism will be discussed later.

3.4.2 Surface Analytical Techniques

Only recently have the chemically specific methods of surface analysis on a microscopic scale been applied to the investigation of optical surface damage mechanisms. Examples of these techniques are Auger spectroscopy, Low Energy Electron Diffraction (LEED), electron energy loss spectroscopy, Laser Ionization Mass Analysis (LIMA), and Scanning Force Microscopy (SFM). The first four of these techniques are electron spectroscopies, which have a limited effectiveness because of surface charging of most insulating optical materials. These methods have mostly been used for large area characterization of surface chemistry and structure. They have not been effective for in situ microscopic characterization before and after laser excitation.

The first applications of LEED to investigate laser ablation were done on GaP by *Kumazaki* et al. [3.48] and on ZnO by *Itoh* [3.49]. For GaP, for example, it was shown that laser excitation at photon energies near the indirect bandgap with a tunable dye laser erased the LEED pattern of the 17×17 reconstructed surface. The threshold fluence for this erasure was correlated with the threshold for neutral Ga emission discussed previously. *Schildbach* and *Hamza* [3.46] performed a similar experiment on the reconstructed (1120) surface of sapphire. When the sapphire surface is prepared by ion bombardment and heated to about 1500 K, it exhibits a 12×4 LEED pattern of a reconstructed surface. Multiple-pulse laser excitation with 10 ns pulses at wavelengths of 1064 nm and 355 nm slowly modifies the characteristics of the LEED pattern. Two types of modification were observed. The first consists of a gradual increase of the crossover value of the incident electron energy at which secondary electron emission balances the charging effect of the primary electron beam. This suggests that the surface conductivity or secondary emission characteristics are altered. Secondly, after a very large number (~ 50000) of pulses at a fluence of over 3.5 J/cm^2, the higher order spots (caused by the reconstruction) in the LEED pattern gradually fade away. This fluence is well below the single-pulse optical-damage threshold of 9 J/cm^2 measured for this surface. Both of these effects were attributed to the observed desorption of Al$^+$ ions, discussed previously. It was suggested that the change in the crossover is caused by the modification of the surface electronic band that is believed to be the reason for the high surface conductivity of sapphire. These large area LEED studies are interesting, because changes in the LEED pattern indicate that the surface modification induced by the laser is not confined to a few small defect sites, but includes a large fraction of the illuminated surface.

Schildbach and *Hamza* [3.46] also performed Auger measurements on sapphire surfaces. These measurements are done by bombarding the surface with an electron beam that promotes electrons from core states in the atoms very near the surface. The Auger decay of the core holes yields secondary electrons with well defined kinetic energies that identify the emitting element and provide bonding information in the form of small energy shifts. No changes in the surface composition due to laser excitation were detected, but it was found that oxygen was lost from the surface due to the primary Auger-electron bombardment. This is just the reverse of what is observed due to laser excitation, where only Al^+ ions are observed to desorb. This illustrates that the surface ablation processes on insulators can be not only elementally selective, but also strongly dependent on the mode of energy deposition onto the surface.

It is important to be able to detect and identify the defect states that may be involved in laser excitation of nominally transparent materials. One method for doing this is Electron Energy Loss Spectroscopy (EELS), in which mono-energetic electrons with energy ~ 200 eV are scattered from the surface, and the scattered electrons are energy resolved so that their energy loss to excitations created on the surface is known. *Schildbach* and *Hamza* obtained the EELS spectrum of 12×4 reconstructed (1120) surface of sapphire [3.46]. Their spectrum of scattered electron flux as a function of energy loss is illustrated in Fig. 3.12. At low sensitivity (lower curve) the interband transition threshold at about 9 eV is somewhat resolution broadened. At higher energy, around 22 eV, is a peak from the volume plasmon. At a $45 \times$ higher sensitivity a peak appears at energies well within the bulk bandgap. This feature is attributed to surface states in bandgap. Although the exact energy distribution of this spectrum is affected by the ~ 1.6 eV (full width at half maximum) resolution broadening of the instrument, it is clear that the energy loss extends from the order of 1 eV up to the bandgap energy. This spans the photon energy range of the fundamental

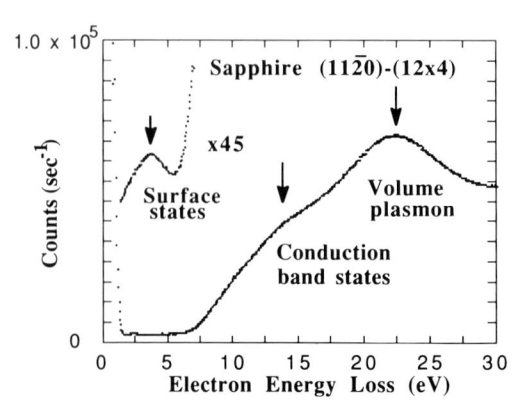

Fig. 3.12. Reflection Electron Energy Loss Spectrum (EELS) for the sapphire samples from the same source as those used for the data in Fig. 3.10. The incident electron energy was 200 eV [3.46]

and third harmonics of the Nd:YAG laser used for the ablation experiments. In fact, the peak of the distribution is at 3.6 eV, near the third-harmonic photon energy. Schildbach and Hamza suggest that this loss peak is due to intrinsic surface states. Unfortunately, it cannot be known from this experiment if these states are filled states within about 3.6 eV of the conduction band, or empty states above the valence band by this same amount. It is possible, however, that the excitations produced by laser absorption into this spectrum are the source of the electronic excitations that lead to desorption of the Al^+ ions.

Electron beam excitation has also provided some valuable evidence concerning the influence of electronic surface excitation on laser ablation. *Dickinson* et al. showed that the irradiation of a sodium trisilicate glass surface by electron beam excitation greatly modifies the threshold and statistical nature of laser ablation [3.50]. After irradiation with 1.5 keV electrons, ablation occurred at a lower fluence, and it was much more reproducible from shot to shot. They also did electron energy loss measurements on the surface before and after the electron irradiation. Following electron irradiation, they observed an electron energy loss feature, similar to the one in the upper trace in Fig. 3.12, at energies below the bulk bandgap. The appearance of this feature was attributed to the production of surface defect states that cause absorption at electron or photon energies less than the bandgap energy. These experiments provide dramatic evidence that such states can have a profound influence on optical damage and laser ablation thresholds.

The development of the Scanning Force Microscope (SFM) has provided the opportunity to investigate submicroscopic surface modification by laser irradiation and the effects of surface roughness and morphology on optical surface damage. In an SFM a probe tip is scanned near a surface and the deflection, or alternatively the force on a cantilever beam supporting the tip, is measured. A two-dimensional plot of the detected signal provides a contour map of the surface that can have atomic scale horizontal and vertical resolution in very favorable circumstances [3.51]. A resolution of less than 10 nm is routinely achievable. Of greatest importance for the investigation of optical damage is the fact that the scanned tip can be removed from the surface during laser excitation and brought back afterwards to scan the same region in order to detect changes made by the irradiation. This experiment was first performed by *Siekhaus* and coworkers, who investigated laser ablation of graphite [3.52], sapphire [3.53], and, most recently, multilayer optical coatings [3.54, 55].

Some representative SFM measurements on a sapphire sample from the same batch used by *Schildbach* and *Hamza* in their work on laser-induced desorption are shown in Fig. 3.13 [3.53]. The round, dark spots are due to optical damage in the form of pits. A majority of these pits are overlapping the polishing scratches that are visible as linear structures, although some damage sites have no apparent physical defects near them. This significant correlation of damage sites with scratches is suggestive of the reduced damage thresholds to be expected from electric-field enhancements from such discontinuities. These images certainly do not yet firmly substantiate this mechanism, but the efficacy

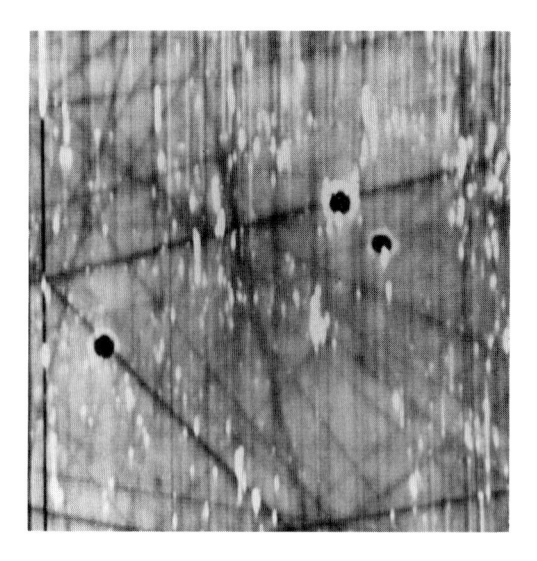

Fig. 3.13. Scanning force microscope image of a polished sapphire surface obtained from the same batch of polished material used for the data of Figs. 3.10 and 12

of the SFM for investigating this aspect of optical damage is clearly established. More extensive measurements for a wider variety of surface conditions for a single material would be very valuable in this regard.

The SFM has also been used by *Staggs* et al. and *Kozlowski* et al. [3.53, 54], and by *Tesar* et al. [3.55], to investigate the effects of laser excitation on multilayer optical coatings. Figure 3.14 shows the changes observed in a multilayer high-reflection coating of about 30 alternating layers of HfO_2 and SiO_2 when it is laser conditioned below its optical damage threshold, as discussed earlier. The surface of the outer SiO_2 layer is coarsened by the laser excitation. There are sizable chunks of the coating (not shown in Fig. 3.14) that are excavated at widely spaced locations. The roughening effect is not understood yet, but the excavation of chunks is consistent with the type of interpretation given earlier for the observation of bursts of particles of very high kinetic energy during conditioning [3.47]. It is now believed that these chunks are growth nodules of coating material that are loosely bound into the coating and are many layers in thickness [3.54]. The images in Fig. 3.14 illustrate one obvious limitation of the SFM for the study of multilayer coatings: it is only sensitive to changes in the surface topography. Nevertheless, the SFM is one of the most powerful new tools for optical damage investigation.

3.4.3 Laser Pump–Probe Measurements

Despite the wealth of information that has been obtained from the experiments discussed so far, they seldom provided any unambiguous indication of the physical processes, such as the basic absorption mechanism of the laser light, that initiate optical surface damage. Recently, a pump–probe experiment was developed that has provided some of this information for bare surfaces and

3. Laser Ablation and Optical Surface Damage   75

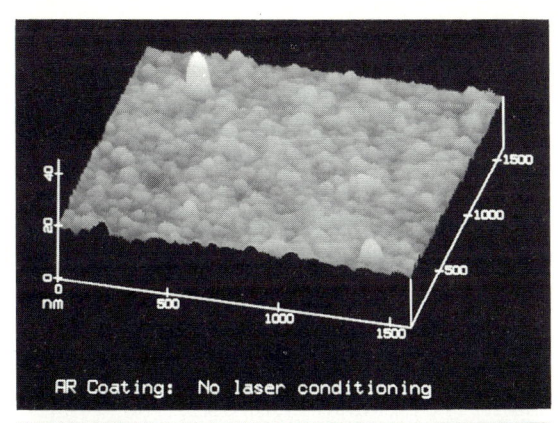

Fig. 3.14. Scanning force microscope images of a HfO_2–SiO_2 multilayer high-reflectivity coating of the type used for Fig. 3.11 before (*above*) and after (*below*) irradiation with laser light at fluence that causes laser conditioning [3.65]

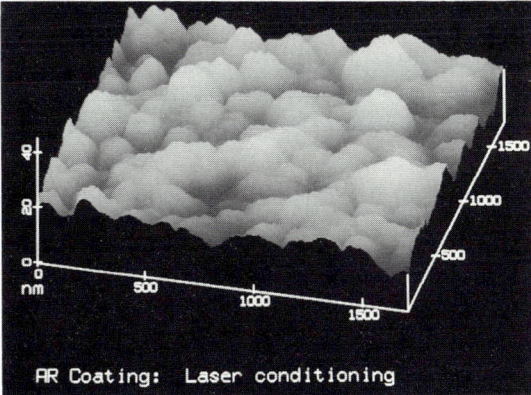

coatings [3.56]. This experiment can reveal whether the laser–surface interaction is linear or nonlinear in the laser intensity and the lifetimes of the excitations (thermal, electronic, acoustic, etc.) produced by this interaction. The idea is to measure the thresholds for optical damage, or for laser ablation, for a single laser pulse and then for a pair of laser pulses separated by a variable time delay. The experimental information consists of the ratios of the single-pulse and double-pulse damage thresholds as a function of time delay [3.56].

In order to interpret this pump–probe experiment it is assumed that the interaction of a single laser pulse with the surface produces some type of excitation with amplitude ϱ that is the primary cause of the surface damage, and that the magnitude of this excitation decays in time after the laser pulse ends. It is further assumed that the excitation must exceed some threshold value ϱ_t in order to cause a damage event above the threshold fluence. When two pulses are applied, the excitation produced by the second pulse adds to the remaining excitation produced by the first pulse. Thus, even if neither pulse exceeds the optical damage threshold fluence the damage threshold may be exceeded by the pair of pulses. These assumptions are certainly true for some simple damage mechanisms, such as linear or nonlinear absorption leading to heating and then to melting, vaporization, or fracture. These assumptions may not be appropriate for such complex phenomena as avalanche ionization, but the experimental results should still be useful, and the model can be altered to fit the experimental observations.

The simplest model that incorporates these assumptions is based on two additional assumptions. Firstly, it is assumed that the rate of energy absorption dE/dt from the laser beam depends only on the laser intensity I as a power law $dE/dt = KI^n$, where n is the order of the absorption process, and K is a constant. The rate of increase of the excitation is assumed to be proportional to dE/dt. In general it is possible that K could vary with time during the laser pulses or the time interval between pulses if there is accumulated surface modification. Secondly, it is assumed that the excitation produced by the first pulse decays exponentially with a time constant τ_e. The excitation density $\varrho(2, \tau)$ immediately after the second pulse can be written as

$$\varrho(2, \tau) = c\left(\frac{F(2, \tau)}{2}\right)^n (1 + e^{-\tau/\tau_e}) , \qquad (3.3)$$

where the first term is the excitation remaining from the first pulse, and the second term is due to the second pulse, $F(2, \tau)/2$ is defined as the fluence of each pulse [the total fluence is then $F(2, \tau)$], and c is a constant proportional to K. The assumption of a unique threshold value of the excitation in order for optical damage to occur means that the value of the excitation should be equal for either single-pulse or two-pulse excitation, so $\varrho_t(1) = cF_t(1)^n = \varrho_t(2, \tau)$. Solving for the ratio of the single-pulse and double-pulse damage fluences, $R(\tau) = F_t(2, \tau)/F_t(1)$ gives the useful result

$$R(\tau) = 2(1 + e^{-\tau/\tau_e})^{-1/n} , \qquad (3.4)$$

In general, a much more complicated relation for $R(\tau)$ might be applicable. Some possibilities are: cooling of a heated region due to thermal conduction or radiation; distributions of relaxation times for different surface regions and defects; distributions of threshold excitations or of values for the constant c. The details of the model are not as important as the basic idea of the experiment itself: assuming that the pulses are identical in intensity and fluence, the ratio of fluence to intensity is increased by a factor of two with two pulses as opposed to one pulse, with the added feature of a time delay. Note that at small time delays $\tau \ll \tau_e$, the ratio R become $2^{(1-1/n)}$, from which n can be determined, independent of whether or not the relaxation is exponential in time. For linear absorption $R(0) = 1$, for two photon absorption $R(0) = 1.414$, etc., and for successively higher orders of nonlinearity, $R(0)$ approaches 2. On the other hand, at sufficiently long time delays $\tau \gg \tau_e$, each pulse must independently exceed the damage threshold, which requires that R approaches 2, regardless of n. The time delay required for R to increase from its value at $\tau = 0$ to $R = 2$ gives a measure of the duration of the excitation, regardless of the specific details of the relaxation process.

This experiment has been performed on several types of bare surfaces and optical coatings. The samples are mounted in a UHV chamber, and a quadrupole mass spectrometer is used to detect neutrals and ions ablated from the sample surface [3.56]. This provides a sensitive indication of when the damage or ablation threshold has been reached, since copious emission of neutrals and ions occurs. This allows a rapid data-taking procedure since microscopic observations are not required after each laser shot. The first application of this experiment was for the polished surface of crystalline ZnS irradiated by picosecond laser pulses at a wavelength of 600 nm [3.56]. ZnS is interesting because it is a good example of a material that exhibits a cumulative damage behavior under multiple pulse excitation. This behavior is illustrated in Fig. 3.15, which shows the emission yield of neutral Zn (and possibility also S_2) as a function of the number of laser pulses for several pulse fluences. Rather abruptly, after several hundred shots above a fluence of 0.4 J/cm^2, optical damage is observed. Below this fluence, many thousands of pulses do not cause damage. The morphology of the damaged regions is that of surface ablation over a large portion of the central region of the laser spot, which is suggestive of ablation by a plasma near the surface [3.56].

The double-pulse threshold experiment for ZnS is illustrative of several aspects of optical damage, and it is worth reviewing the results for this material in some detail. The ratio $R(\tau)$ for multiple pulse damage of ZnS is plotted as a function of τ in Fig. 3.16. At $\tau < t_p$, where t_p is the pulse duration, the two pulses overlap temporally, and interference fringes form on the surface, which doubles the intensity in the regions of the fringe maxima. It is easily shown that $R = 0.5$ in this case, independent of n or τ_e, since the redistribution of fluence causes the

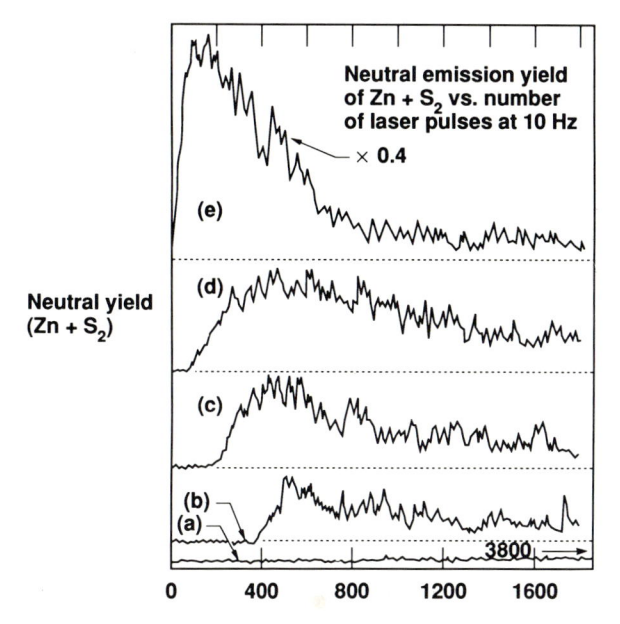

Fig. 3.15. Accumulated optical damage studied by measurement of the yield of neutral Zn and S_2 emission induced by repetitive picosecond laser pulse excitation at a 10 Hz rate. Laser wavelength is 600 nm. Approximate fluences in J/cm^2 are: (*a*) 0.38; (*b*) 0.40; (*c*) 0.43; (*d*) 0.56; (*e*) 1.35

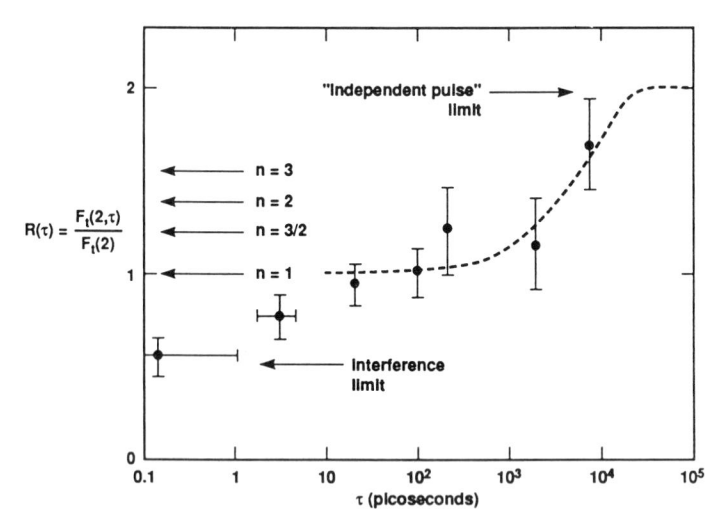

Fig. 3.16. The fluence ratio $R(\tau)$ defined in the text as a function of time delay between two picosecond pulses for a Zn single-crystal surface. The dashed curve is a fit to (3.4) with $n = 1$ and $\tau_e = 5$ ns

single-pulse damage fluence to be exceeded in the fringe maxima at a double-pulse fluence half as large as the single-pulse damage fluence. This value ($R = 0.5$) is observed for values of τ up to a few ps, approximately the autocorrelation time of the pulses, and this provides some assurance that the experimental approach is valid. For 10 ps $< \tau <$ 1 ns, two orders of magnitude in time delay, $R = 1$, which shows that linear absorption causes the surface modification that leads to optical damage after several hundreds of pulses. The dashed curve in Fig. 3.16 is a fit to (3.4) with a lifetime of 5 ns for the excitation created by the laser. This lifetime has been attributed to thermal conduction away from microscopically heated regions [3.56].

The indication that linear absorption is the relevant laser–surface interaction for ZnS is at first sight somewhat surprising. The laser wavelength is well above the threshold for bulk two-photon absorption [3.57], and the maximum intensity of these picosecond pulses is over 200 GW/cm^2 at the damage threshold. In fact, two-photon absorption depletes the laser-beam intensity to a few percent of its initial value after passing through the 0.5 mm thickness of this sample. Using the reported values of the two-photon absorption coefficient [3.58], it can be shown that about 200 J/cm^3 is deposited in the sample near the entrance surface by two photon absorption. This is only sufficient to raise the sample temperature by about 200 degrees. More importantly, however, it leads to a carrier density of about 3×10^{20} cm^{-3}, well above the estimated densities that are required to "seed" an electron avalanche, assuming that the laser intensity is high enough to initiate it. Apparently, the avalanche ionization rate is not fast enough to give significant carrier multiplication during the time duration of the laser pulse. This is perhaps reasonable in light of the avalanche rates for SiO_2 shown in Fig. 3.5. At 200 GW/cm^2, the avalanche rate would be well below 10^9 s^{-1}, much smaller than the inverse of the laser pulse duration (1 ps). Of course, ZnS has a much smaller bandgap than SiO_2, and higher avalanche rates would be expected, but it is likely that intensities close to 1 TW/cm^2 would be required for significant avalanche ionization during a picosecond pulse.

The large free-carrier densities resulting from two photon absorption in ZnS point to absorption by the free carriers as another potential damage mechanism for this crystal. If the rate of energy loss per electron is assumed to be the same as the calculated values for SiO_2 given in Fig. 3.4, and the carrier density is 3×10^{20} cm^{-3}, it is easily shown that the free carrier absorption deposits about 1500 J/cm^3 into the sample at the optical damage threshold. This is considerably larger than the direct energy deposited by the two-photon absorption, and it could be sufficient to account for the accumulation effect leading to optical damage. The conclusion reached from the double-pulse experiment, that linear absorption is the cause of optical damage shows, however, that free-carrier heating is not the cause of the damage. These results for ZnS are important because they show that multiphoton absorption is not necessarily a significant mechanism for optical surface damage even where rather large two-photon absorption cross sections are involved.

In order to account for the results of the double-pulse experiment, as well as other, less direct evidence, it has been suggested that optical surface damage of ZnS is caused by heating of linearly absorbing surface regions, possibly ones that are to some degree thermally isolated from the surrounding medium [3.31, 37, 56]. What could be the nature of these regions? It is possible that they are concentrations of linearly absorbing point defects, such as F-centers, which absorb over a large spectral range in bulk ZnS [3.56]. The cumulative modification of the surface that leads to multiple-pulse optical damage might be caused by electronic or thermal desorption that modifies the composition of the near-surface region and increases the absorption on successive shots.

This pump probe experiment has been done with 1.06 μm laser pulses of subpicosecond and 80 ps duration on sapphire [3.59, 60]. Linear-or low-order absorption was indicated by the ratio $R \approx 1$ at short time delays. This is consistent with the suggestion of *Schildbach* and *Hamza* that laser ablation of Al$^+$ from sapphire is due to absorption caused by surface states. Quite different lifetimes were deduced for the excitation causing ablation in these experiments, which were done with pulse durations differing by a factor of about 100, however, and this behavior remains unexplained [3.60]. This illustrates the importance of the pulse duration on optical damage mechanisms.

Investigations of optical coating damage mechanisms using the double-pulse experiment were performed for single-layer, quarter-wave (≈ 0.18 μm thick) coatings of SiO_2 and HfO_2 using 80 ps pulses [3.61]. The data for the fused-silica coatings showed clearly that linear absorption is the damage mechanism, and the dependence of the ratio $R(\tau)$ on time delay showed that the lifetime of the excitation caused by this absorption is extremely long, about 35 ns. The damage morphology consisted of submicron pits scattered randomly throughout the irradiated area. The long decay time of the excitation suggests that it is heating of microscopic regions. The mean size r of the heated regions should be related to the relaxation time and the thermal diffusivity of the medium by $r = (D\tau_e)^{1/2}$. If the value of D for bulk fused silica is used, this yields a mean size of $r = 0.1$ μm for the heated regions. Since the laser-pulse duration is negligible compared with the thermal relaxation time, the measured threshold fluence (10 J/cm^2) can be used together with the inferred size of the absorbing regions to estimate the absorption coefficient in the absorbing regions. Assuming that a temperature rise of about 2000°C is sufficient to cause damage, the required absorption is $\alpha \sim 300$ cm^{-1}. It was speculated that these regions consist of nonstoichiometric material SiO$_x$, with x < 2 [3.61]. These estimates illustrate the important point that absorbing regions of submicron dimensions can cause thermally initiated optical damage in coatings if they are merely "gray" not "black".

Double-pulse experiments on a HfO$_2$ coating also gave noteworthy results. In the course of these measurements it was found that Hf neutrals were ejected from the coating with high kinetic energies at fluences far below the single-pulse damage threshold at a given fluence and then disappeared. This behavior is similar to that discussed earlier for the HfO$_2$–SiO$_2$ multilayers during laser conditioning [3.47], and this emission is very likely caused by the same mechanism responsible for the laser conditioning of these multilayers. The ratio $R(\tau)$ increased very gradually from near unity at short pulse delays (≈ 0.1 ns) up towards 2 at the longest delays of several nanoseconds. This behavior was interpreted as resulting from linear absorption at defects with a distribution of relaxation times [3.61]. These films are partially crystalline, and this behavior might result from variations in thermal conduction rates due to grain boundaries and columnar growth.

It is apparent from the work discussed above that linear absorption is an important optical damage mechanism for several types of bare surfaces and coatings. The only material studied by the double-pulse experiment for which the optical damage mechanism is clearly due to a nonlinear laser–surface interaction is a well-polished bare surface of BK-7 optical glass [3.56]. The ratio $R(\tau)$ for this case measured with laser pulses of picosecond duration is shown in Fig. 3.17. At the shortest pulse delays, $R(\tau)$ is not far below 2, and it barely changes with τ. It was speculated that the optical damage in this case might be a true case of "optical breakdown" due to an electron avalanche [3.56]. the peak power at the damage threshold was, however, only about 500 GW/cm^2, which is just barely at the onset of a significant ionization rate according to the calculations for SiO$_2$ in Fig. 3.5. Additional studies of this material with longer pulse durations would be very useful.

Another class of pump–probe experiments that have demonstrated considerable potential for investigations of optical surface damage are based on photothermal displacive or refractive phenomena. These experiments involve detection of laser-heated surface regions or of heated gas near the surface by the

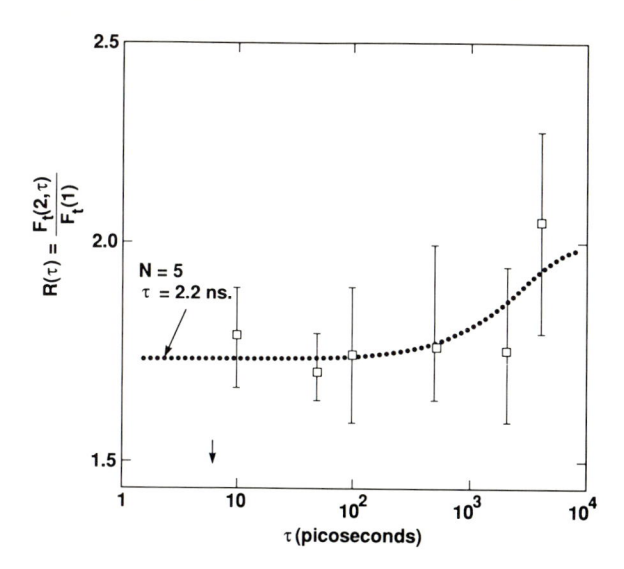

Fig. 3.17. The fluence ratio $R(\tau)$ for an optically polished borosilicate (BK-7) glass surface. The dashed line is a fit to (3.4). Note the change in scale from Fig. 3.16

displacement of a probe laser beam. There are two general variants of this experiment. In the first variant, the "bulging" of a locally heated surface region due to thermal expansion deflects the weak probe beam that is focused on the surface near the center of an intense pump beam. This is called PhotoThermal Deflection (PTD). Alternatively, the surface reflectivity may be temperature dependent, so PhotoThermal Reflectivity (PTR) and its time dependence may be used to detect the heating and thermal diffusion in the heated region. Using the first of these approaches, *Abate* et al. [3.62] observed heating at specific surface sites below the optical damage threshold with spatial resolutions as good as a few microns. Unfortunately, there is not always a good correlation established (or attempted) between these heated regions and the regions at which optical damage is initiated at higher fluences. *Wu* et al. have recently used PDT and PTR to investigate heating of thin films and multilayer optical coatings and to measure their thermal conduction properties. They employed modulated pump and probe techniques as a function of modulation frequency and obtained thermal conductivity data for several oxide and fluoride films [3.63] and diamond films [3.64]. They also did an experiment in which the time-delay of the photothermal deflection was used to detect thermal waves and determined the heating of each coating layer [3.65]. The observed time delay is due to the time required for thermal conduction of the heat to the outer coating layer. Specific types of defects were artificially introduced into a coating, and photothermal microscopy was used to detect them [3.66]. These experiments have considerable potential for investigating absorbing defects in surfaces and coatings.

In the second type of experiment, the probe beam is passed near to the surface and is deflected by the change in the refractive index cause by thermal expansion or by shock waves emitted from the heated surface. *Petzold* et al.

[3.67], and *Matthias* et al. [3.68] have done such experiments on cleaved surfaces of several fluoride crystals. They observe photothermal deflection above a threshold fluence and calculate from the measured deflection the energy deposited into the surface and plasma by the laser. Above this threshold, they observe a rapid increase in the deflection, which they fit to a power law in the laser intensity. It is not clear if this power-law emission occurs above or below the damage threshold, since they measure the average reflectivity of the irradiated region as an indication of damage, so localized damage spots might not be detected. As is the case for the measurements shown in Fig. 3.9, there is the question of whether the inferred multiphoton process occurs in the gas phase or in the solid.

3.5 Concluding Remarks

In the past five or six years, new experimental methods have provided new insights into the fundamental mechanisms of optical surface damage and surface ablation. These methods have been most effective when several of them were applied to investigate a single material or coating, as has been done, for example, for ZnS, GaP, Al_2O_3, the fluorite structure crystals, CaF_2 and BaF_2, and HfO_2–SiO_2 coatings, several of which were used as examples in this article. There is now an opportunity to apply these techniques to determine the important surface and coating properties that must be controlled in order to raise damage thresholds and improve and performance and reliability of optical materials. It is particularly important to continue to apply the full suite of available techniques to individual materials and coatings in order to thoroughly understand their behavior when irradiated by high-power lasers. In addition, this very productive experimental situation could benefit considerably from theoretical work on the basic mechanism of laser–surface interactions and laser ablation mechanisms.

Acknowledgements. This work was supported by the Division of Materials Sciences, Office of Basic Energy Sciences of the U. S. Department of Energy and by Lawrence Livermore National Laboratory under Contract No. W-7405-ENG-48.

References

3.1   R.F. Wood: *Laser Damage in Optical Materials* (Adam Hilger, Bristol 1986)
3.2   Reviews of laser-materials interactions, optical damage, and related topics are published in a topical issue of Opt. Eng **28** (1989)
3.3   E. Matthias, R.W. Dreyfus: In *Photoacoustic, Photothermal, and Photochemical Processes at Surfaces and in Thin Films*, ed. by P. Hess Springer Topics Cur. Phys. Vol. 47 (Springer, Berlin, Heidelberg 1989) p. 89

3. Laser Ablation and Optical Surface Damage   83

3.4   L.L. Chase, A.W. Hamza, H.W. W. Lee: In *Laser Ablation: Mechanisms and Applications*, Springer Lect. Notes Phys., Vol. 389, ed. by J.C. Miller, R.F. Haglund (Springer, Berlin, Heidelberg 1991) p. 193
3.5   F. Rainer, R.M. Brusasco, J.H. Campbell, F.P. Demarco, R.P. Gonzales, M.R. Koslowski, F.P. Milanovich, A.J. Morgan, M.S. Scrivener, M.C. Staggs, I.M. Thomas, S.P. Velsko, C.R. Wolfe: In *Laser Induced Damage in Optical Materials 1989*. NIST Special Publication No. 801. SPIE **1438**,74 (1990)
3.6   S.R. Foltyn, L.J. Jolin: In *Laser Induced Damage in Optical Materials 1986*. NBS Special Publication No. 752, 182 (1987)
3.7   P.K. Bandyopadhyay, L.D. Merkle: J. Appl. Phys. **63**, 1392 (1988)
3.8   W.H. Lowdermilk, D. Milam: IEEE J. QE-**17**, 1888 (1981)
3.9   C.R. Wolfe, M.R. Koslowski, J.H. Campbell, R. Rainer, A.J. Morgan, R.P. Gonzales: In *Laser Induced Damage in Optical Materials 1989*. NIST Special Publication No. 801. SPIE **1438**, 360 (1990)
3.10 J.E. Rothenberg, R. Kelly: Nucl. Instrum Methods Phys. Res. **229**, 291 (1984)
3.11 J.E. Rothenberg, R. Kelly; Nucl. Instrum Methods Phys. Res. **B7/8**, 755 (1985)
3.12 C.R. Giuliano: In *Laser Induced Damage in Optical Materials 1972*. NBS Special Publication No. 372, 49 (1972)
3.13 J. Kardach, A.F. Stewart, A.H. Guenther: In *Laser Induced Damage in Optical Materials 1982*. NBS Special Publication No. 669, 164 (1982)
3.14 M. Sparks, D.L. Mills, R. Warren, T. Holstein, A.A. Maradudin, L.J. Sham, E. Loh, Jr., D.F. King: Phys. Rev. **B24**, 3519 (1981)
3.15 D. Arnold, E. Cartier, M.V. Fischetti: In *Laser Induced Damage in Optical Materials 1990*. SPIE **1441**, 478 (1990)
3.16 A.S. Epifanov: Sov. Phys. JETP **40**, 897 (1974)
3.17 A.S. Epifanov, A.A. Manenkov, A.M. Prokhorov: Sov. Phys. JETP **43**, 377 (1979)
3.18 B.G. Gorshkov, A.S. Epifanov, A.A. Manenkov: Sov. Phys. JETP **49**, 309 (1979)
3.19 S.C. Jones, P. Braunlich, R. Thomas Casper, X.A. Shen, P. Kelly: Opt. Eng. **28**, 1039 (1989)
3.20 D. Arnold, E. Cartier, D.J. Dimaria: Phys. Rev. B **44**, 1477 (1992)
3.21 D. Arnold, E. Cartier: In *Laser Induced Damage in Optical Materials 1992*. SPIE **1848**, 424 (1992)
3.22 R.W. Hopper, D.R. Uhlmann: J. Appl. Phys. **41**, 4023 (1970)
3.23 M. Sparks, C.J. Duthler: J. Appl. Phys. **44**, 3038 (1973)
3.24 J.H. Pitts: In *Laser Induced Damage in Optical Materials 1985*. NBS Special Publication No. 746, 236 (1987)
3.25 M.F. Kodunov, A.A. Manenkov, I.L. Pokotillo: sov. J. Quant. Electron. **18**, 345 (1988)
3.26 N. Bloembergen: Appl. Opt. **12**, 661 (1973)
3.27 N. Bloembergen: IEEE J. QE-**10**, 375 (1974)
3.28 R.A. House, J.R. Bettis, A.H. Guenther: IEEE J. QE-**5**, 361 (1978)
3.29 D.L. Rousseau, G.E. Leroi, W.E. Falconer: J. Appl. Phys. **39**, 3328 (1968)
3.30 A. Schmid, P. Braunlich, P.K. Rol: Phys. Rev. Lett. **35**, 1382 (1975)
3.31 L.L. Chase, L.K. Smith: In *Laser Induced Damage in Optical Materials 1987*. NBS Special Publication No. 756, 165 (1988)
3.32 J.T. Dickinson: In *Adhesive Chemistry Developments and Trends*, ed. by L.H. Lee (Plenum, New York 1984) p. 97
      J.T. Dickinson, L.C. Jensen, M.R. McKay: J. Vac. Sci. Technol. A **5**, 1162 (1987) and references therein
3.33 M.L. Knotek: Rep. Prog. Phys. **47**, 1499 (1984)
3.34 M.L. Knotek, P.J. Feibelman: Phys. Rev. Lett. **40**, 964 (1978)
3.35 K. Tanimura, N. Itoh: Nucl. Instrum. and Methods Phys. Res. B **33**, 815 (1988)
3.36 N. Itoh, T. Nakayama: Phys. Lett. **92A**, 471 (1982)
      Y. Nakai, K. Hattori, N. Itoh: Appl. Phys. Lett. **56**, 1980 (1990)
      N. Itoh, K. Hattori, Y. Nakai, J. Kansaki, A. Okano, R.F. Haglund: In *Laser Abalation: Mechanisms and Applications*, ed. by J.C. Miller, R.F. Haglund, Springer Lect. Notes Phys., Vol. 389 (Springer, Berlin, Heidelberg 1991) p. 213

3.37 H.F. Arlinghaus, W.F. Calaway, C.E. Young, M.J. Pellin, D.M. Gruen, L.L. Chase: J. Appl. Phys. **65**, 281 (1989)
3.38 T. Nakayama, H. Ichikawa, N. Itoh: Surf. Sci. Lett. **123**, L693 (1982)
3.39 M. Okigawa, T. Nakayama, N. Itoh: Nucl. Instrum. Methods Phys. Res. B **9**, 60 (1985)
3.40 W.J. Siekhaus, J.H. Kinney, D. Milam, L.L. Chase; Appl. Phys. A **39**, 163 (1986)
3.41 E. Matthias, M.B. Nielson, J. Reif, A. Rosen, E. Westin: J. Vac. Sci. Technol. B **5**, 1415 (1987)
3.42 J. Reif, H. Fallgren, W.E. Cooke, E. Matthias: Appl. Phys. Lett. **49**, 770 (1986)
3.43 J. Reif, H. Fallgren, H.B. Nielson, E. Matthias: Appl. Phys. Lett. **49**, 930 (1986)
3.44 J. Reif: Opt. Eng. **28**, 1122 (1989)
3.45 R.W. Dreyfus, F.A. McDonald, R.J. Vongutfeld: J. Vac. Sci. Technol. B **5**, 1521 (1987)
3.46 M.A. Schildbach, A.V. Hamza: Phys. Rev. B **5**, 1521 (1987)
3.47 M.A. Schildbach, L.L. Chase, A. V. Hamza: In *Laser Induced Damage in Optical Materials 1990*. SPIE **1441**, 287 (1991)
3.48 Y. Kumazaki, Y. Nakai, N. Itoh: Phys. Rev. Lett. **59**, 2883 (1987)
3.49 N. Itoh: Nucl. Instrum. Methods. Phys. Res. B **27**, 155 (1987)
3.50 J.T. Dickinson, S.C. Langford, L.C. Jensen, P.A. Eschbach, L.R. Pederson, D.R. Baer: J. Appl. Phys. **68**, 1831 (1990)
J.T. Dickinson, S.C. Langford, L.C. Jensen: In *Laser Ablation: Mechanisms and Applications*, ed. by J.C. Miller, R.F. Haglund, Springer Lect. Notes Phys., Vol. 389 (Springer, Berlin, Heidelberg 1991) p. 301
3.51 H.K. Wickramasinghe (ed.): *Scanned Probe Microscopy*, AIP Conf. Proc., Vol. 241 (American Institute of Physics, New York 1991)
3.52 R.J. Tench, M.A. Schildbach, M. Balooch, W.J. Siekhaus, A.A. Tesar: In *Scanned Probe Micoscopy*, ed. by H.K. Wickramasinghe, AIP Conf. Proc. Vol. 241 (American Institute of Physics, New York 1991) p. 490
3.53 M.C. Staggs, M. Balooch, M.R. Kozlowski, W.J. Siekhaus: *Laser Induced Damage in Optical Materials 1991*. SPIE **1624**, 375 (1992)
3.54 M.R. Kozlowski, M. Staggs, M. Balooch, R. Tench, W.J. Siekhaus: SPIE **1556**, 181 (1992)
3.55 A.A. Tesar, M. Balooch, K.W. Schotts, W.J. Siekhaus: In *Laser Induced Damage in Optical Materials 1990*. SPIE **1441**, 228 (1991)
3.56 L.L. Chase, H.W.H. Lee, R.S. Hughes: Appl. Phys. Lett. **57**, 443 (1990)
3.57 M. Sheik-Bahae, A.A. Said, T.H. Wei, D.J. Hagan, E.W. VanStryland: IEEE J. QE **26**, 760 (1990)
3.58 E.W. VanStryland, L.L. Chase: In *CRC Handbook of Laser Technology, Suppl. 2: Optical Materials*, ed. by M.J. Weber, to be published
3.59 A.V. Hamza, R.S. Hughes, L.L. Chase, H.W.H. Lee: J. Vac. Sci. Technol. B **10**, 228 (1992)
3.60 A.V. Hamza, R.S. Hughes, L.L. Chase, H.W.H. Lee: In *Laser Induced Damage in Optical Materials 1991*. SPIE **1624,** 429 (1992)
3.61 L.L. Chase, A.V. Hamza, H.W.H. Lee: J. Appl. Phys. **71**, 1204 (1992)
3.62 J.A. Abate, A. Schmid, M.J. Guardalben, D.J. Smith, S.D. Jacobs: In *Laser Induced Damage in Optical Materials 1983*. NBS Special Publication No. 688, 385 (1983)
3.63 Z.L. Wu, M. Reichling, H. Gronbach, Z.X. Fan, D. Schaefer, E. Matthias: In *Laser Induced Damage in Optical Materials 1991*. SPIE **1624**, 33 (1992)
3.64 Z.L. Wu, H. Gronbeck, X. Su, Z.X. Fan: In *Laser Induced Damage in Optical Materials 1991*. SPIE **1624** (1992)
3.65 Z.L. Wu, Z.X. Fan, D. Schaefer: In *Laser Induced Damage in Optical Materials 1991*. SPIE **1624** (1992)
3.66 Z.L. Wu, M. Reichling, E. Welsch, D. Schaefer, Z.X. Fan, E. Matthias: In *Laser Induced Damage in Optical Materials 1991*. SPIE **1624** (1992)
3.67 S. Petzoldt, A.P. Elg, M. Reichling, J. Reif, E. Matthias: Appl. Phys. Lett. **53**, 2005 (1988)
3.68 E. Matthias, S. Petzoldt, A.P. Elg, P.J. West, J. Reif: In *Laser Induced Damage in Optical Materials 1988*. NIST Spcial Publication No. 756, 217 (1988)

4. Pulsed-Laser Deposition of High-Temperature Superconducting Thin Films

T.V. Venkatesan

With 17 Figures

The development of High-Temperature Superconducting (HTS) thin-film technology has been significantly enhanced by the introduction of Pulsed-Laser Deposition (PLD). This technique excels in the deposition of multi-element compounds such as the high temperature superconductors and a variety of other compatible materials such as dielectrics and conducting layers. In this paper I will review the advantages and problems with this technique and some of the most exciting recent breakthroughs that have occurred in the high-temperature superconducting thin-film arena based on the pulsed-laser film-deposition technique.

4.1 Advantages of Pulsed-Laser Deposition

Ever since the discovery of the laser a few decades ago the potential for pulsed-laser deposition of thin films has remained unexploited [4.1]. Despite the sustained pioneering work at Rockwell in the area of laser deposition [4.2], it took the development of the high-temperature superconductors to fully realize the potential of this technique. In the early work on pulsed-laser deposition of high-temperature superconductors [4.3] it was first demonstrated that the composition of rather complex multi-elementary materials can be reproduced in the film under appropriate conditions of laser-energy density and deposition angle with respect to the target surface normal [4.4]. These features were significantly unique to the pulsed laser deposition process and with the recipe for ·making in situ, crystalline films of proper stoichiometry known [4.5], the feverish pace of the research activity in the field of high-temperature superconductors and the versatility and ease of the technique has significantly enhanced its popularity in the research community.

A simple schematic of the pulsed laser deposition system is shown in Fig. 4.1 where an excimer laser pulse (wavelength of 248 nm, pulse width of 30 ns) of energy density of a few J/cm^2 is incident on a stoichiometric target of the film of interest. The evaporated material is predominantly ejected in the forward direction [4.4] and under suitable background pressures of oxygen (100 mTorr) and substrate temperatures (of around 750 °C), high quality superconducting films are grown in situ.

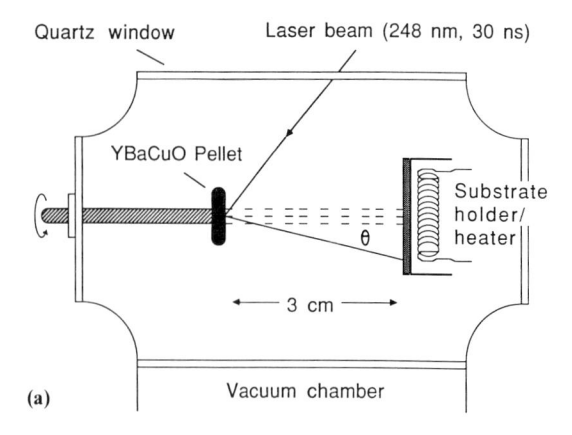

Fig. 4.1. A schematic of a simple pulsed laser deposition system (**a**) and (**b**) a PLD system in operation (courtesy of Neocera, Inc., College Park, MD)

The features of the laser deposition process that make it so unique are the following:

1. Rather complex multi-element materials can be deposited easily provided a single-phase, homogeneous target can be fabricated. The complexity of the deposition process is translated to the relatively easier process of fabricating a high-quality target.
2. The chamber-pressure, the target–substrate distance, the target orientation with respect to the laser beam, etc. are significantly decoupled enabling a greater degree of freedom in the deposition system design. The target is decoupled from the substrate in the sense that a small target can be used to deposit a film over a fairly large-area substrate with the appropriate scanning schemes.

3. The efficiency of the target use is superior compared to any other technique since a predominant amount of the evaporated material is forward directed and can be collected with a high degree of efficiency. For example, in a production environment over 100 YBCO films (of thicknesses ranging from 3000–4000 Å) on 1 cm^2 substrates have been fabricated from a one inch target of 0.25 inch thickness with a majority of the target unused [4.6]. The cost of raw materials in a production environment may become significant and, particularly for toxic elements there is a further advantage in minimizing the spread of the contaminants [4.7].
4. The forward-directed evaporant in this process lends itself to novel substrate handling schemes such as the recessed "black body" heater, to be described later, which may not be possible with most sputtering techniques.
5. The fabrication of multi-layers is fairly straightforward with rapid substitution of targets into the path of the laser beam. Most materials have a fairly common range of evaporation parameters. Hence, it is relatively easy to design automated systems capable of sophisticated thin film structures. Some of the most complex integrated high-T_c components were demonstrated recently using pulsed-laser deposition [4.8].

4.2 Materials Base

The concerted effort in developing in situ processes for the deposition of high-temperature superconductors and the associated dielectric and buffer layers has given us the confidence to extend this technique to a variety of other materials.

As a matter of fact, a number of applications of this technique for the fabrication of electro-optic materials [4.9], ferro-electric materials [4.10] etc. have already been demonstrated. One of the important spin-offs has been the cross fertilization of materials such as the use of YBCO films as room temperature electrodes for hetero-epitaxial ferro-electric layers [4.11]. It is becoming fairly clear that, as the technology of epitaxial, multi-element materials evolves, the unique capabilities of pulsed-laser deposition are going to make this an indispensable technology. Besides successful deposition of superconductors such as the YBCO, LSCO [4.12], BSSCO [4.13], TBCCO [4.14], NdCeCuO (NCCO) [4.15], and BaKBiO (BKBO) [4.16] this technique has been extended to the fabrication of a variety of dielectric layers such as LaAlO$_3$ [4.17], SrAlTaO$_3$ [4.18], YSZ [4.19, 4.20], CeO$_2$ [4.21] and metallic layers such as SrRuO$_3$ [4.9, A.22], and LaNiO$_3$ [4.23], to name only a few. In the rest of this paper we will address some of the forefront thin film experiments that have been performed utilizing pulsed laser deposition. We will end with issues relating to the commercial scaleup of this process and a discussion of problems that have been solved as well as those that need solution.

4.3 Laser-Beam–Target Interaction

4.3.1 Target Texturing

One of the important problems recognized fairly early was the texturing of the target by the incident laser beam which inevitably makes a non-normal incidence angle on the target. Due to the shadowing effects, as the target is irradiated, surface features are amplified via cone formation as illustrated in Fig. 4.2 [4.24]. The cone formation leads to three problems; a drop in the deposition rate with time [4.24], an increase in the incidence of particle formation on the films and, last but not the least, a shift in the angle of the emitted evaporant towards the laser by as much as 20 degrees. Since the composition and the film thickness is optimum at the peak of the emitted plume of materials from the surface a shift of the plume direction is a serious problem for the production of high-quality thin films.

A number of schemes have been devised to overcome this problem, one of which is to produce a line focus on the target and to position the beam symmetrically about the center of the rotating target. Then the beam can be rastered up and down to prevent the formation of a depression at the center of the target [4.25]. The symmetric positioning of the beam about the target center automatically reverses the direction of the laser beam with respect to the cones formed, thus mitigating any shadowing effects.

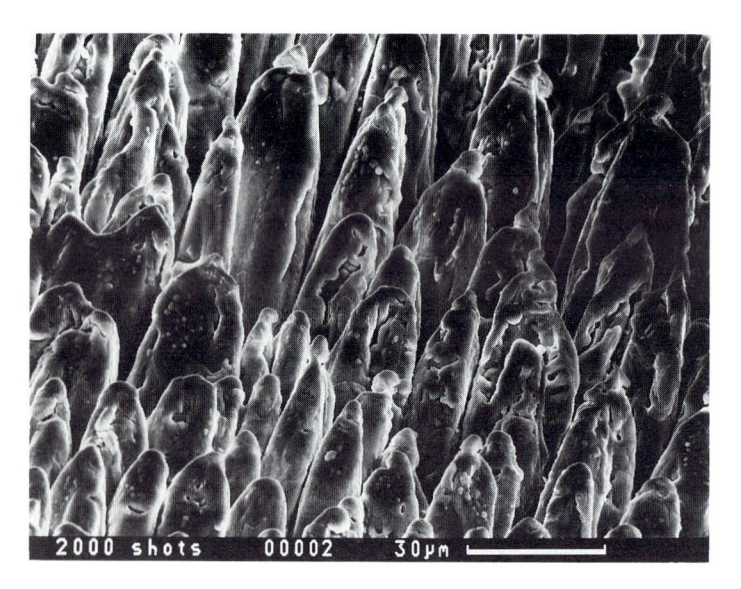

Fig. 4.2. Scanning electron microscope image of cones formed on the surface of the target after laser irradiation due to shadowing effects [4.24]

4. Pulsed-Laser Deposition of High-Temperature Superconducting Thin Films   89

4.3.2 Particle Deposition

One of the disadvantages of the laser-deposition technique with respect to most other techniques which was recognized early was the problem of particle deposition in the films. The origin of this problem is multi-faceted, but is primarily related to processes that will increase the surface features on the target. The idea is that at times large objects connected to the target surface by narrow regions could have their anchor evaporate earlier thereby forcibly ejecting the particulate matter towards the substrate. There have been two approaches to solve this problem. The first is the passive approach where the production of the particulate matter has been minimized by working with

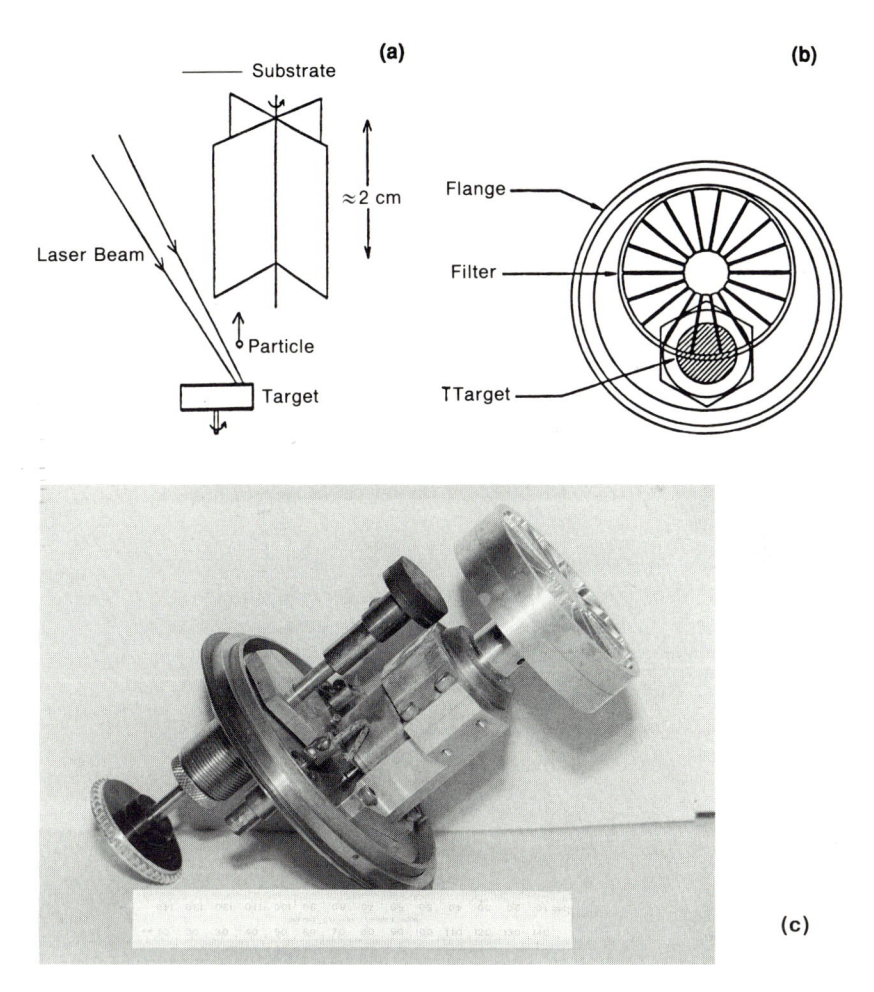

Fig. 4.3. (a) Schematic of velocity filter operation, (b) top view of velocity filter, and (c) side-view photo of filter [4.26]

targets of high homogeneity and density. It is very important for the target to have a single homogeneous phase. Further, it is common practice to polish the target periodically to make the surface free of any features that could potentially be dislodged during the evaporation process. The problem of particulate deposition has been significantly diminished by this approach. For dielectric materials this problem seems to be less significant, for reasons which are not fully understood. Another idea has been the active approach where by using a velocity filter one is able to stop the slowly moving massive particulate matter while transmitting the fast atomic and molecular evaporants.

The results of such experiments done at Neocera, Inc. are shown in Fig. 4.3, where the velocity filter is shown with the schematic as well as the original machined component [4.26]. The vane-like structure eliminates the slow moving components and allows only those species that have velocities exceeding nfl, where n is the number of vanes, f is the number of revolutions per second and l is the length of the vane. The results of the experiment are shown in Fig. 4.4, where a dramatic decrease in the particle density in the film is seen when the velocity filter is used in the laser plume. In this particular example the filter was operated at a maximum speed of 3300 rev/min which resulted in an order of magnitude decrease in the particle density in the films. Today, with the availability of high speed motors that are vacuum compatible even more efficient velocity filters have been made. As a result, when high density,

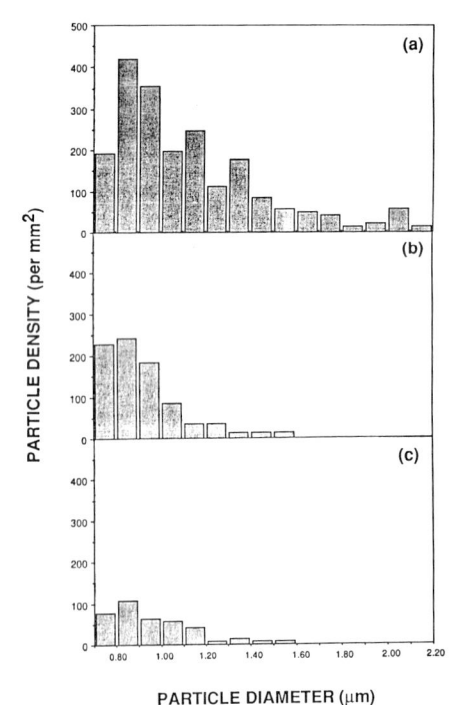

Fig. 4.4. Histogram of particle distribution vs diameter for different voltages applied to the filter motor. (**a**) 0.1 V (36 rpm), (**b**) 8.0 V (2400 rpm), and (**c**) 11 V (3300 rpm) [4.26]

homogeneous targets are not available, the incorporation of velocity filters in the laser plume may be a viable option. The idea of a synchronous chopper has also been pursued [4.27] which would entail a lower demand on the motor assembly used to turn the chopper. A novel idea for particle reduction by using a second intersecting beam to further evaporate the initial evaporants from the target has been demonstrated by *Koren* et al. [4.28]. But it is not clear whether this technique can compete with a mechanical chopper in terms of simplicity and costs.

4.4 Dynamics of the Laser-Produced Plume

The dynamics of the evaporated material at the target surface has been extensively studied by a number of groups [4.29–31]. The predominant species observed tend to be atomic in nature with simple compounds formed by fragmentation of the original molecule [4.32]. *Geohegan* and *Mashburn* [4.33] observed that the ion fraction increased rapidly with increasing laser energy density though it is unclear as to the role of ions in the deposition process. One of the earliest studies of the evaporants established the existance of two distinct components in the laser-produced fragments at the target surface [4.4] (Fig. 4.5). One of the components with a weaker angular dependence was identified as a non-stoichiometric thermal evaporation and the other component, highly peaked in the forward direction, was identified as the stoichiometric component produced in a nonlinear evaporation process such as shock-wave formation or a highly excited dense plasma. The relative ratio of material emitted in the two different components depends upon the laser energy density and, at lower energy densities (typically for YBCO below 0.9 J/cm², for 248 nm

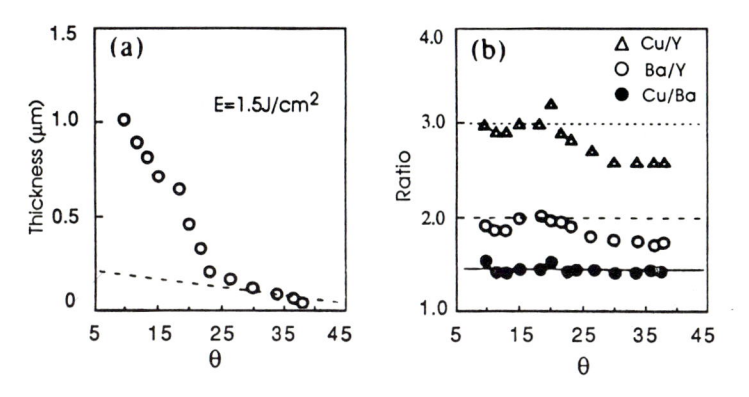

Fig. 4.5. (a) The film thickness vs deposition angle showing two distinct components in the laser evaporation process. The dashed line is a simple cos Θ fit. (b) The composition as measured by RBS as a function of the deposition angle showing the lack of stoichiometry for the non-forward directed component [4.4]

and 30 ns laser pulses), the thermal evaporation component dominates as a result of which the composition of the deposited film deviates from the proper stoichiometry. At higher laser energy densities the propensity for deposition of laser-produced debris increases as a result of which an optimum energy density is required for the production of high-quality films with proper surface morphology.

The origin of these different evaporative components is a subject of intense study. Direct time of flight measurements have been performed as a function of the emission angle and laser energy density by *Lynds* et al. [4.34] which indicate high non-thermal energies for the particles emitted in the forward direction. *Kelly* and *Dreyfus* [4.35] have modelled the evaporation of high-density evaporants predicting forward peaking purely from multiple collisional effects. While this model can explain the phenomenon, it is inadequate to explain the observation of very energetic non-thermal species [4.33, 34] in the evaporants from the surface. Recently, ultrafast photographic techniques have been used by *Gupta* et al. to observe the temporal evolution of the fragments and the resultant shock wave formed in an ambient gas of oxygen [4.36], as shown in Fig. 4.6. The results of estimates of the surface temperature of the target close to the melt temperature ($\approx$ 1400 K) as well as the power density threshold of 5.5×10^7 W/cm^2, within a factor of two of estimates of thermal evaporation thresholds of YBCO, tend to suggest the role of evaporation in initiating the initial ejection of materials. As the gas density increases above the target surface inverse bremsstrahlung absorption of the laser energy can heat the plasma to temperatures higher than the surface of the target consistent with an observed increase in the expansion front velocity at the end of the laser pulse. Using a laser-induced fluorescence technique the evaporated species were shown to consist of a twin

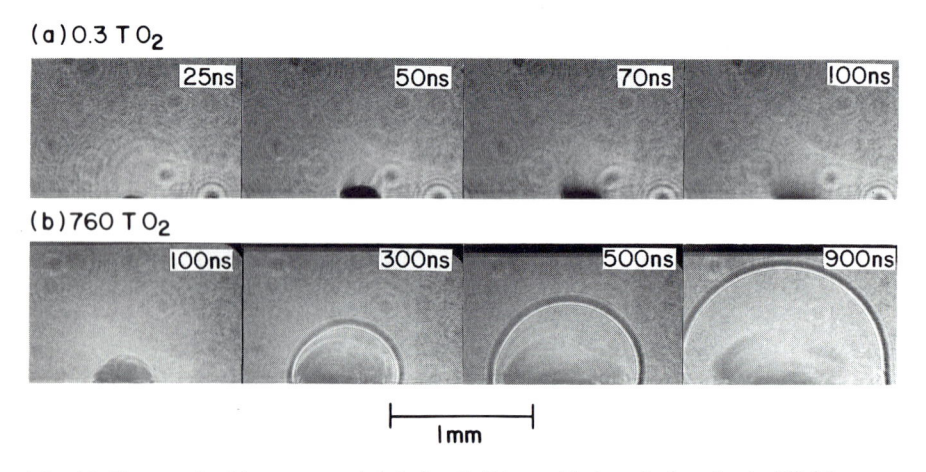

(a) 0.3 T O$_2$

(b) 760 T O$_2$

Fig. 4.6. Time-resolved images recorded during KrF-laser ablation of a low-density YBCO target using a single pulse for photographing each frame. The oxygen background pressure is 0.3 Torr in (**a**) and 760 Torr in (**b**). The top edge of the YBCO target is at the bottom of each frame. The recorded time is from the start of the laser pulse [4.36]

velocity distribution consisting of a slow and fast component by *Okada* et al. [4.37]. Upon introduction of oxygen into the system the slow component was quenched indicating a modification of the velocity distribution of the species by the ambient gas. This observation is consistent with the earlier observation of two different angular distributions in the evaporant species and the stoichiometric, forward-directed component is likely to be the faster component as well since the composition is preserved in the films independent of the oxygen pressure.

Besides dynamical measurements, spectroscopic techniques have been used by many groups to characterize the evaporated materials and their resultant interactions with the background gas [4.38–40]. *Wu* et al. [4.40] have shown that, as the oxygen pressure is increased in the system, the fraction of the YO molecular optical emission steadily increases, with a concomitant decrease in the optical emission of the atomic lines. These results tend to suggest optical spectroscopic probes as a monitoring scheme for the gas-phase reactions inside a PLD system. Spatially resolved optical measurements have enabled *Zheng* et al. [4.41] and *Girault* et al. [4.42] to measure the velocity distribution of the emitted material with non-thermal velocities of the order of 10^4 m/s, consistent with observations by laser-induced fluorescence.

4.5 Evaporant–Substrate Interaction

Due to the high pressures involved during the deposition process (> 50 mTorr) conventional surface analysis tools such as particle-based spectroscopic tools do not work well at all. Recently, two different approaches have been adopted to overcome these problems. The first, by *Kanai* et al. [4.43], dubbed laser MBE, utilizes a low pressure deposition scheme with NO_2 as the background gas at a pressure of 10^{-5} mTorr with a Reflection High Energy Electron Diffraction (RHEED) system to perform the in situ observation. The system works very well for molecular systems that do not require large oxygen partial pressures to be stable (unlike YBCO). Using this technique the group has fabricated films of (Ca, Sr) CuO_2 "infinite-layer" superconductors with evidence for superconductivity at 80K and above [4.44]. RHEED oscillations were indeed observed in this system indicating a layer-by-layer growth.

Another novel technique to enable in situ surface analysis has been the use of a pulsed oxygen source with pulse lengths on the order of 0.5ms. This provides a large instantaneous pressure during the deposition of the species, which lasts for a period of say 20–30 ms, without raising the chamber pressure above, say, 10^{-5} Torr. Using this source, in fact, it was shown [4.45] that a high oxygen flux is needed on the YBCO surface during film growth and a delay of as little as 100 ms in the oxygen arrival relative to the evaporant at the surface is sufficient to result in poor films at the substrate. These results are illustrated in Fig. 4.7 where the resistivity vs temperature characteristics are shown for a variety of

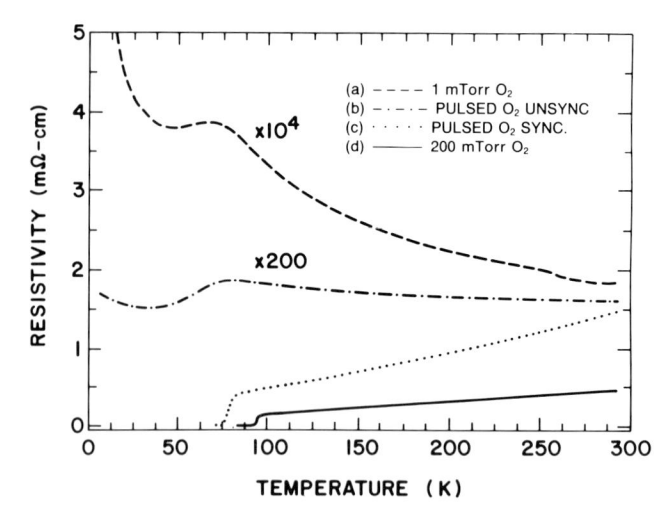

Fig. 4.7. Resistivity vs temperature results of films deposited under different conditions and source gases using the pulsed valve. (a) Ablation fragments are deposited 250 ms before arrival of the O_2 jet at the substrate, target–substrate distance (X) = 8.0 cm (*dashed*); (b) fragments are deposited during presence of O_2 at the substrate with X = 4.5 cm (*dashed-dotted*); (c) same as in (b), with X = 8.0 cm; and (d) using N_2O source gas with the triggering sequence as in (b) and (c), X = 8.0 cm. The O_2 flux used during depositions of films (a-c) is $\approx 2 \times 10^9/cm^2 s$; the N_2O flux for film (d) is about a factor of 2 lower than the O_2 flux [4.45]

arrival times of the evaporants and the gas pulse. Using the pulsed oxygen source *Chern* et al [4.46] have demonstrated layer-by-layer growth of LSCO and YBCO films using in situ RHEED and though the quality of these films are inferior to the films made at higher oxygen pressures the technique could be further developed to be more effective. Clearly, these techniques are leading pulsed-laser deposition techniques into arenas that have been the niche of UHV techniques such as MBE. In situ conductivity measurements during the PLD process have been very useful in yielding information on the oxygen and annealing kinetics in the film [4.47].

4.6 Frontiers of High-Temperature Superconducting Thin-Film Research

4.6.1 Epitaxial Multilayers

In situ deposition of very high-quality crystalline epitaxial films have become very routine and the challenges seem to be in the area of fabrication of a variety of useful heterostructures. In analogy with the GaAs/AlAs hetero-structure which has been the driving force behind most group III–V-based devices, in the YBCO-based superconductors the doping of Pr in the Y site (PBCO) yields a similar lattice-matched materials system with which a variety of heterostructures have been demonstrated. YBCO/PBCO/YBCO sandwich structures have been utilized to fabricate SNS devices with good weak–link properties [4.48, 49] exhibiting potential for microwave detection. YBCO/SrTiO$_3$ structures have been utilized to fabricate multilayer device structures that integrate a Superconducting Quantum Interference Device (SQUID) and a flux transformer (Figs. 4.8 a, b) [4.8, 50] and these integrated structures have been used to record the first magnetocardiogram at 77 K using SQUID technology (Fig. 4.8 c). The most significant demonstration of the potential of the current deposition processes has been the fabrication of superlattices of YBCO/PBCO where the layer thicknesses approach unit cell dimensions (Fig. 4.9). In such structures remarkable effects have been observed such as the reduction in the transition temperatures with the unit cell thickness (Fig. 4.10) which has been interpreted as a reduction in the interlayer coupling effects [4.51–53], or as due to hole filling by the Pr layers [4.54] and so on. Since by using Z-contrast microscopy *Pennycook* et al. [4.55] have demonstrated the atomic layer abruptness of the superlattices prepared by PLD, the concept of synthesizing novel layered

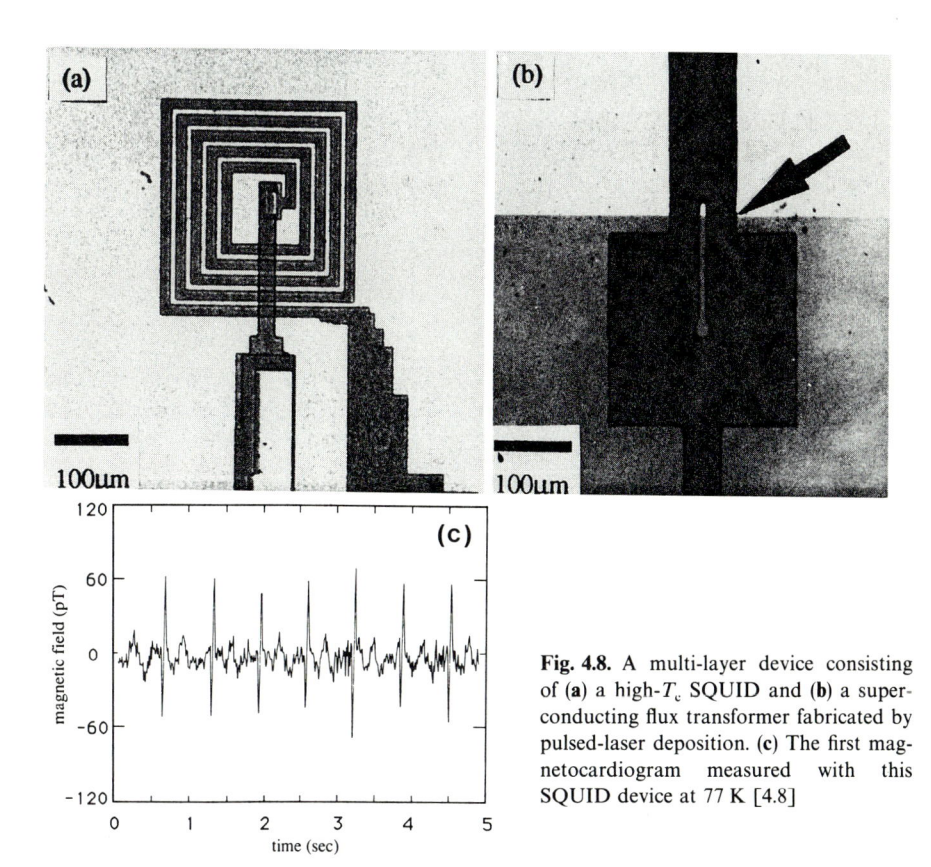

Fig. 4.8. A multi-layer device consisting of (a) a high-T_c SQUID and (b) a superconducting flux transformer fabricated by pulsed-laser deposition. (c) The first magnetocardiogram measured with this SQUID device at 77 K [4.8]

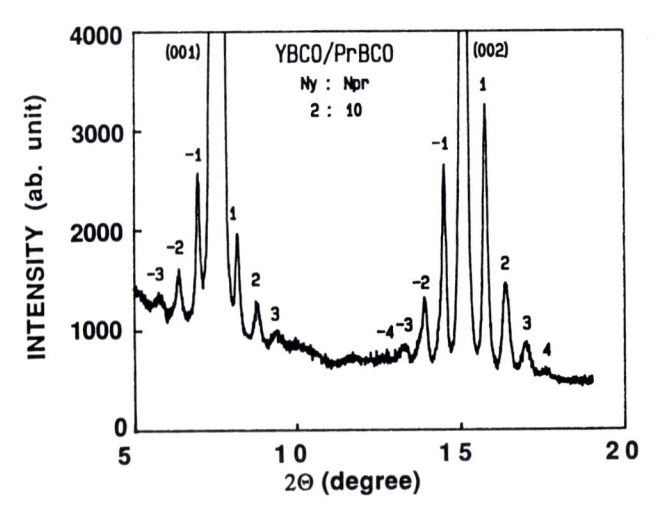

Fig. 4.9. X-ray Θ–2Θ spectra for a YBCO/PrBCO superlattice with alternating periods of 2 YBCO unit cells and 10 PrBCO unit cells. Observe the well-defined satellite peaks next to the (00L) lines due to the superlattice modulation [4.51]

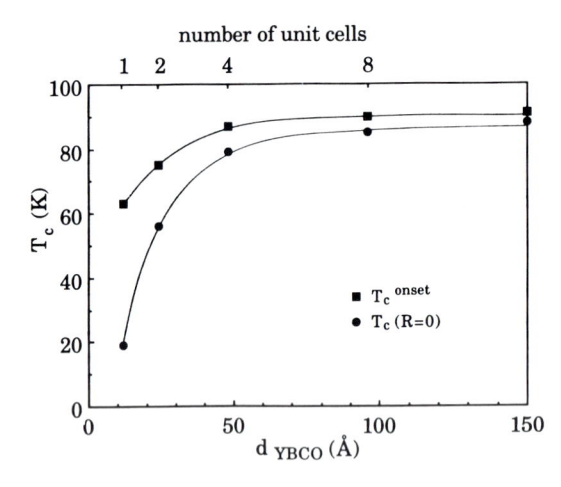

Fig. 4.10. The transition temperature of a superlattice sample where the superconducting layers of YBCO approach unit cell dimensions [4.51]

compounds using PLD is becoming an avenue to be exploited both for technological and basic studies of these layered compounds.

4.6.2 Work on Ultrathin Films

A recent exciting area of development has been the observation of field effects in high-temperature superconductors which may usher in an era of a number of novel three-terminal devices in these materials [4.56, 57]. Using YBCO films of thicknesses approaching unit cell dimensions (1–5 unit cells = 12–60 Å) and an

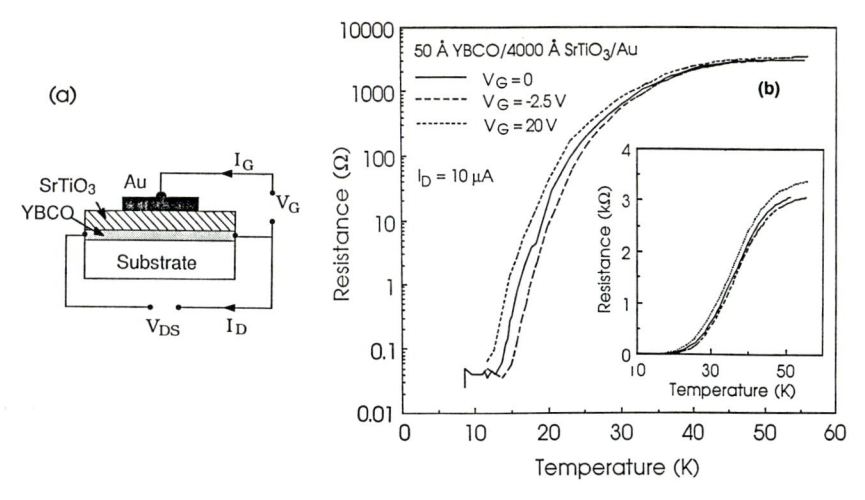

Fig. 4.11. (a) Schematic of the three-terminal device; (b) resistance vs temperature curves for a three-terminal field-effect device consisting of a 50 Å YBCO layer as the channel and a 4000 Å SrTiO$_3$ layer as the gate dielectric for different gate voltages [4.56]

epitaxial SrTiO$_3$ gate electrode the superconductivity in the channel was modulated by a gate electrode. The device structure is shown in Fig. 4.11a and this structure is fabricated in situ by a pulsed-laser deposition technique by substituting the appropriate target in the laser beam. In Fig. 4.11b is shown the resistance versus temperature curves for a 50 Å YBCO channel layer for different values of the gate voltage. One sees a clear shift in the transition temperature of the YBCO layer and at a temperature of, say, 15 K, the ratio of the resistivities for a voltage of 20 V to −2.5 V is about 30. By optimization of the film properties these ratios may exceed 1000. The field effect in YBCO may open the possibilities for high-current switches that have a low resistance in the on-state so that power dissipation is minimized by such devices. Both examples presented for the superconductors illustrate the advanced state of the thin films capability in this materials system. It is ironical that before the advent of high-temperature superconductors one of the most exciting frontiers of superconductivity research was in the area of superlattices. The current state of the art for high-temperature superconductors in this area of superlattices is comparable to what is being accomplished in low-temperature superconductors.

4.6.3 Control of Phase and Crystallinity in Thin-Film Form

The thermodynamics of crystallization and phase formation may be significantly different for films and bulk materials. Using pulsed laser deposition it is fairly easy to produce a variety of compounds that may be otherwise hard to synthesize in bulk form. The early example of the work of *Kanai* et al. [4.43] in synthesizing the so-called infinite-layer superlattices is an example where to synthesize this compound in bulk requires fairly high pressures during the annealing process. However, using PLD, these films have been made at very low pressures in a vacuum chamber. Similarly, recently films of TbBaCuO [4.59] have been prepared by PLD process with transition temperatures of 90 K and these materials are not easily synthesized in bulk form. In fact, using the segmented-target technique and a variety of stable sub-oxides as segments one can attempt making a variety of novel phases in thin-film form and this will be a growing area of materials synthesis in the future.

One of the bottlenecks in the advancement of the HTS thin films based technology is the requirement for epitaxy between the film and the substrate. Particularly for high critical current density applications a minimum of large-angle grain boundaries are desired in the film. A film like YBCO grows c-axis textured on virtually any substrate but the absence of orientation locking, when the lattice mismatch is too large or the substrate is polycrystalline, leads to the formation of large-angle in-plane grain boundaries which result in poor film properties [4.60]. A recent experiment by *Iijima* et al. [4.61] involving an Ion Beam Assisted Deposition (IBAD) of YSZ films on a polycrystalline hastalloy substrate has opened up a new avenue of research which could make the extension of high-quality films to a variety of other substrates a reality. The YBCO films made on these oriented YSZ films on the hastalloy substrates exhibit critical-current densities at 77 K in excess of 10^5 A/cm^2 in zero field and only an order of magnitude drop at 8 T, which requires a significant reduction of large-angle grain boundaries. The ion beam incident at a grazing angle tends to orientation lock the crystals in this direction. The origin of this orientation locking on a microscopic level would indeed be an interesting study and could help us control the film-growth processes even more on substrates that do not have a compatible lattice match. The incorporation of an ion beam in a PLD system is an illustration of the ease of including process enhancements to this deposition system and the experimental results of *Iijima* et al. [4.61] is truly a remarkable advancement in the field of HTS thin films.

4.7 Scaling-up to Larger Areas

From the very inception of laser deposition the technique was thought to be incompatible with large areas on account of the directional nature of the laser-produced plume. One could not have misjudged this technology any further. The deposition process must be envisioned as the directional production of a high density of evaporants which could be swept over a surface of arbitrary area by the movement of the target or the substrate. Some of the early experiments on the possible deposition rate have given a major shot-in-the-arm for this technique. Using a 30 W excimer laser it has been demonstrated by the Los Alamos group that deposition of good-quality films may be possible at high rates [4.62]. In Fig. 4.12 the resistance versus temperature curve is shown for an YBCO film deposited at a rate of 150 Å/s.

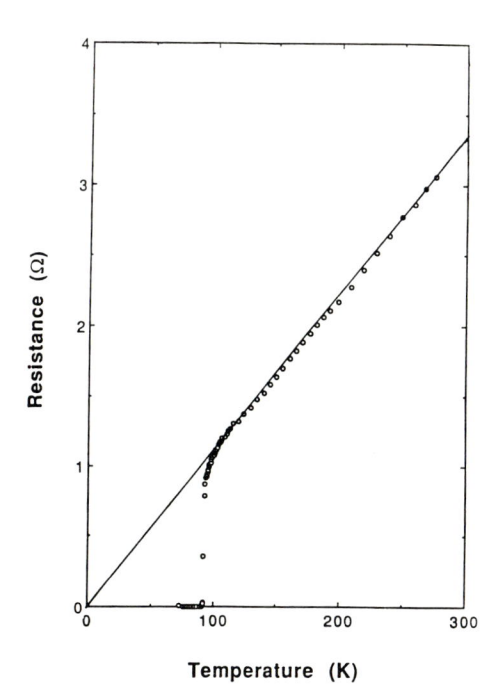

Fig. 4.12. R vs T curve for a YBCO film deposited at a rate of 150 Å/s [4.62]

The resistivity curve clearly shows a film of very high T_c, ΔT_c and quality, as exemplified by the linear extrapolation of the resistivity through the origin. However, when one looks at the defect density in the film as by ion channeling, a different picture emerges. The data shown in Fig. 4.13 indicate the minimum yield (roughly proportional to percentage of disordered atoms in the near surface region) increases with the deposition rate implying that the surface has a fairly long relaxation time (of the order of several msecs) during the deposition process. Of course, by scanning the substrate the negative effect of the rapid deposition rate can be easily mitigated and surprisingly, for properties such as critical currents, the deposition rate does not seem to have any adverse effects what so ever as seen by the critical current density data in both zero (Fig. 4.14) and finite magnetic fields (Fig. 4.15) [4.63]. On a different note, the pulsed laser deposition enables the deposition of a monolayer of material in times of the order of a few milliseconds and hence may be a unique experimental analog of producing the initial state of a system modelled by molecular dynamic simulation. This could lead to a detailed understanding of the surface crystallization phenomenon, provided an appropriate surface probe could be used.

Based on these data one concludes that a deposition rate of 600 Å · cm²/s is possible with a laser of 30 W power. Lambda Physik is developing an experimental laser of 750 W power the use of which will then extrapolate to a deposition rate of 1.5 μm · cm²/s which illustrates the awesome potential of this technique. Such a high deposition rate may be very difficult to achieve by

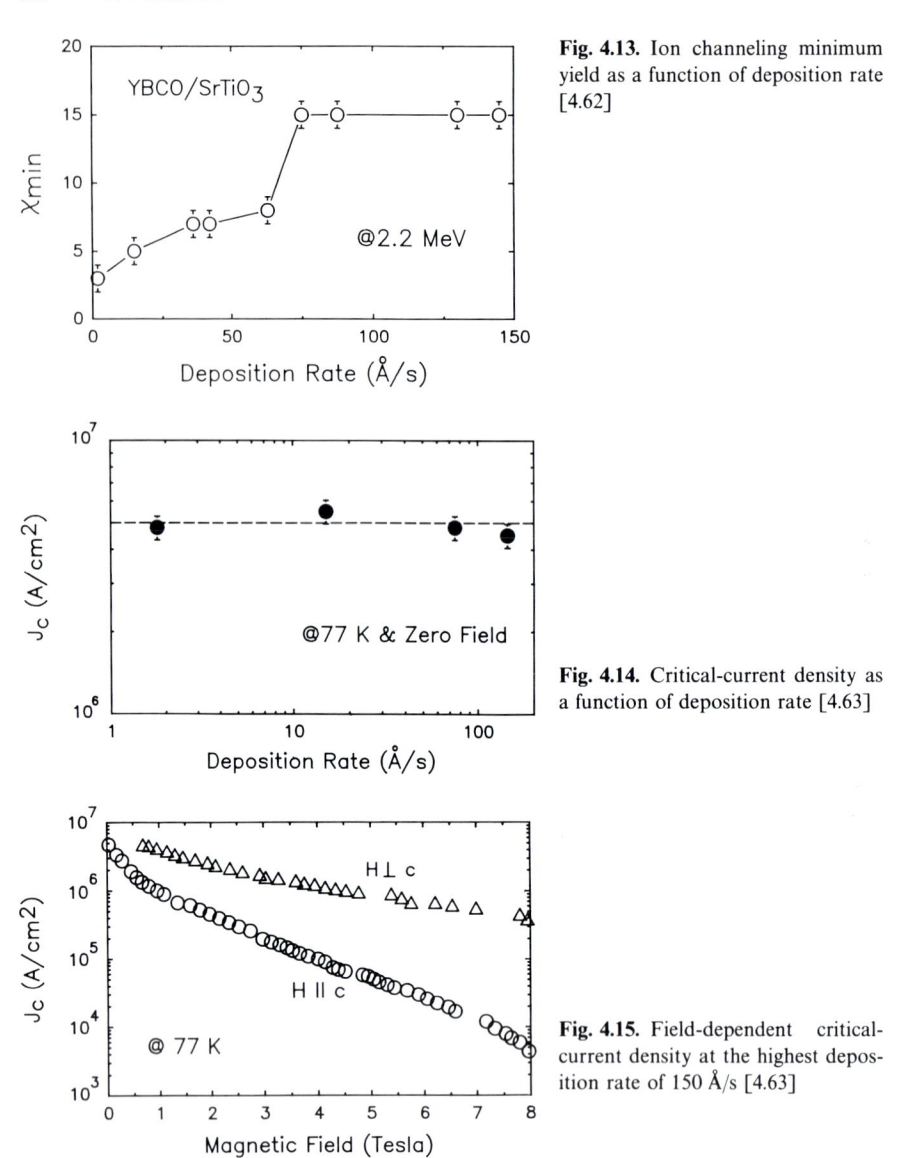

Fig. 4.13. Ion channeling minimum yield as a function of deposition rate [4.62]

Fig. 4.14. Critical-current density as a function of deposition rate [4.63]

Fig. 4.15. Field-dependent critical-current density at the highest deposition rate of 150 Å/s [4.63]

sputtering techniques since the sputtering power cannot be indefinitely increased due to space-charge limitations in the plasma. In both cases, at high deposition rates, the heating of the target is a serious problem and must be solved by novel techniques for heat removal. At the other end of the spectrum, the PLD process may better resemble the slow deposition rate of an MBE process by the utilization of a very low energy laser beam (on the order of μJ per pulse), focussed to produce energy densities of a few Joules per cm^2 and used at a very high repetition rate (kHz or more). The particle kinetics at the surface of the target may be considerably different owing to the difference in the collision kinetics and this may be a regime worth exploring.

By scanning or rotating the substrate, large area deposition with good homogeneity has been achieved over substrates as large as three inches [4.63]. By a simple rotation of the substrate mounted off-axis with respect to the laser produced plume, two inch diameter $LaAlO_3$ substrates have been coated [4.64, 65] with excellent uniformity as shown by the closeness of the transition temperature as in Fig. 4.16 and the uniformity of the microwave properties, as

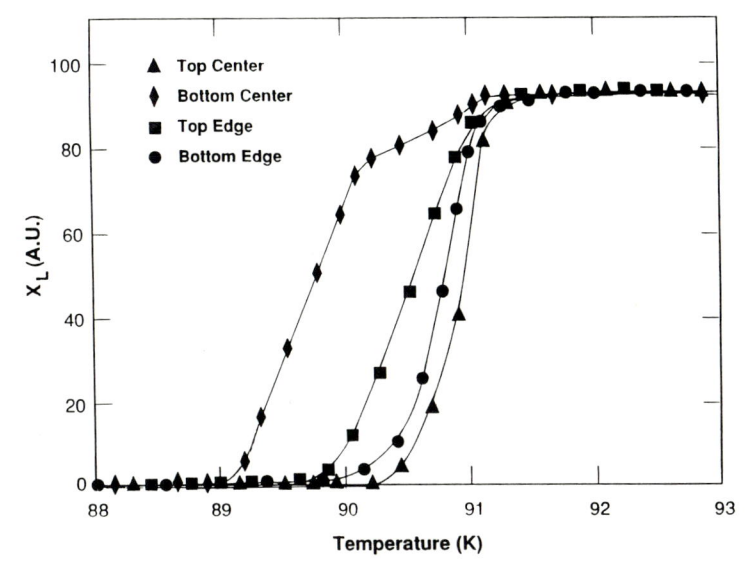

Fig. 4.16. Transition temperature for the center and edge of a two-sided 5 cm coating measured by the dynamic impedance technique; the single-coil instrument allowed independent determination of T_c for each side [4.65]

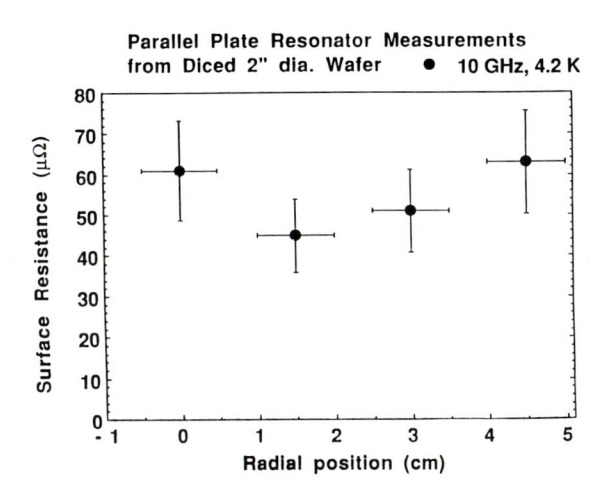

Fig. 4.17. Surface resistance measured at 10 GHz by a parallel-plate resonator technique for two cm² pairs of samples cut from a coated 5 cm wafer. The surface resistance is plotted as a function of position on the wafer [4.65]

seen in Fig. 4.17. The scaling up to large areas by scanning of the target would still necessitate proper preablation and manipulations of the optics to ensure uniform film deposition, as has been pointed out by *Davis* et al. [4.66].

4.8 Future Directions

4.8.1 Component Development

The most serious problem with pulsed-laser deposition is generic to all high-T_c superconductor deposition systems and it is the high deposition temperatures required to make films of very high quality. Despite a number of claims about low deposition temperatures [4.67, 68], based on resistive evaluation of the films, the production of films with low microwave surface resistance or high critical-current densities generally requires substrate surface temperatures approaching 700–750 °C. At these temperatures radiative loss of heat can be substantial and, as a result, maintaining temperature homogeneity and uniformity over space and time is a non-trivial matter. Even the measurement of the temperature is a formidable challenge. Under the conventional heater arrangements involving flat-plate heaters the problems arise from the fact that the sample usually requires some form of a conductive adhesive to hold it on to the heater block, and this process, besides being cumbersome, precludes the possibility of double-side coatings. Even for substrates heated by radiative techniques, the emissivity of the surface changes with the deposition of the film on the substrate which results in a time-varying temperature on the substrate surface. The ultimate solution is the construction of a heater where the substrate is held inside a nearly perfect black body except for an entrance hole through which the laser evaporated plume could be brought in. A perfect "black-body" heater is the ultimate solution and the directional nature of the laser deposition would be a major advantage for these future developments.

The holy grail of high-temperature superconducting thin-film deposition has been the fabrication of films at temperatures approaching semiconductor processing temperatures of around 500 °C. Of all the techniques around today, laser deposition has one of the most important features needed for such a possibility. The highly energetic nature of the evaporants produced at the target by the laser (energies of 50–100 eV), if controlled appropriately at the substrate surface, could result in an effective increase of the surface temperature locally over volumes comparable to the surface diffusion dimensions. The use of pressures of 100 mTorr tends to slow down the species but by incorporating reactive species such as ozone at reduced chamber pressures one may be able to utilize the energy of the evaporants beneficially.

4.8.2 System Issues

One of the advantages of a laser deposition system in a commercial setup or even in an extensive laboratory situation is the ability to time-share a laser among a number of deposition chambers. The actual time involved in the deposition of films may be a small fraction of the cool-down time for the samples in which case the use of multiple chambers with the laser beam steered from one chamber to another may be a prudent option and this further mitigates the capital cost of the laser.

The ideal laser system for the deposition of films is likely to have a pulse energy of about 800 mJ at a repetition rate of 200 Hz, with a beam homogeneity, and a pulse-to-pulse energy stability of better than 10%. The cost of an excimer laser increases nonlinearly with the pulse energy while the differential cost for increase in repetition rate is significantly lower [4.69]. Hence, the improvements in laser technology ought to be in the area of increasing the power output through increase in repetition rate rather than through increase of energy per pulse. As we had discussed earlier, it may be worthwhile to explore the deposition regimes where a very low energy (microjoule/pulse), high repetition rate (tens of kHz) laser is used so that the instantaneous deposition rate may be comparable to what is achieved in MBE-type processes.

The recent experiments of *Kanai* et al. [4.43] tend to suggest the use of laser–target combinations to replace Knudsen cells in MBE systems for special applications which gives an idea of the expanding role for lasers in other deposition systems. In the laboratory environment the pulsed laser is used in the patterning of the superconducting films which suggests the possibility of a multipurpose chamber where the in situ deposition and projection patterning [4.70] of the sample may be ultimately possible.

For in situ multilayer fabrication, laser deposition has natural advantages. Many of the materials used such as YBCO (superconductor), $SrTiO_3$ (dielectric) and platinum (metal) can all be deposited fairly efficiently under virtually the same conditions of laser energy density and substrate temperature for the growth of epitaxial heterostructures. In fact, by the use of a segmented target and by simple rotation of the target superlattices of YBCO/PrBCO layers have been formed where the layer thicknesses have been controlled to the precision of one unit cell dimension [4.50, 51]. This immediately suggests the possibility of applying this technology for the fabrication of multilayers for optical interference coatings or for the ever-challenging X-ray optics – a field whose importance to the technology of micro-lithography cannot be underestimated.

What we are seeing is the tip of an iceberg in terms of the flexibility in the R&D and the commercial environment provided by the PLD process. When one couples the progress made in the field of projection patterning, laser-induced chemistry for the deposition of both dielectric and metallic layers, and novel techniques for laser-induced localized diffusion and doping of species in surface layers [4.70], it is not hard to imagine fairly complex in situ thin-film and device-processing systems, where a number of laser-assisted processes could be coupled to produce novel hetero-structure-based devices. Pulsed-laser deposition is a technology at a very rapid-growth stage with many more niches to conquer.

4.9 Summary

The pulsed-laser deposition technique has made a significant impact in these early days of development of the HTS thin-film research and technology. While the laser deposition is versatile and is ideal for an exploratory environment, the technique has the potential to be scaled-up and hence would be an ideal system to be developed for commercial applications.

Acknowledgements. I would like to thank a number of my colleagues whose contribution has enriched these technologies and brought this to such an exciting stage. In particular, I would like to acknowledge X.D. Wu, X.X. Xi, Q. Li, R. Muenchausen, S. Foltyn, S. Harshavardhan, A. Pique, A. Gupta, and F. Wellstood for help with this manuscript.

References

4.1 J.F. Ready: *Effects of High Power Laser Radiation* (Academic, New York 1971)
4.2 J.P. Cheung, H. Sankur: CRC Crit. Rev. Solid State Mater. Sci. **15**, 63 (1988)
4.3 D. Dijkkamp, T. Venkatesan, X.D. Wu, S.A. Shaheen, N. Jisrawi, Y.H. Min-Lee, W.L. McLean, M. Croft: Appl. Phys. Lett. **51**, 619 (1987)
    X.D. Wu, D. Dijkkamp, S.B. Ogale, A. Inam, E.W. Chase, P.F. Micelli, C.C. Chang, J.M. Tarascon, T. Venkatesan: Appl. Phys. Lett. **51**, 861 (1987)
4.4 T. Venkatesan, X.D. Wu, A. Inam, J.B. Wachtman: Appl. Phys. Lett. **52**, 1193 (1988)
4.5 A. Inam, M.S. Hegde, X.D. Wu, T. Venkatesan, P. England, P.F. Miceli, E.W. Chase, C.C. Chang, J.M. Tarascon, J.B. Wachtman: Appl. Phys. Lett. **53**, 908 (1988)
4.6 M. Mathur: Neocera, Inc. (College Park, MD), private communication
4.7 At STI, Inc. (Santa Barbara, CA) pulsed-laser deposition is used to produce their commercial Thallium based superconducting thin films. The films are produced primarily by a post-annealing technique, but from the point of view of preserving stoichiometry and minimizing contamination the technique has been particularly unique
4.8 Using pulsed-laser deposition a fifteen-layer integrated SQUID and a flux-transformer coil have been successfully fabricated on a single chip at Conductus, Inc. (Sunnyvale, CA) in a collaboration with the University of California at Berkeley.
    A.H. Miklich, F.C. Wellstood, J.J. Kingston, J. Clarke, M.S. Clclough, K. Char, G. Zaharchuck: Nature **352**, 482 (1991)
4.9 S. Yilmaz, T. Venkatesan, R. Gerhard-Multhaupt: Appl. Phys. Lett. **58**, 2479 (1991)
4.10 R. Ramesh, K. Luther, B. Wilkens, D.L. Hart, E. Wang, J.M. Tarascon, A. Inam, X.D. Wu, T. Venkatesan: Appl. Phys. Lett. **57**, 1505 (1990)
4.11 R. Ramesh, A. Inam, B. Wilkens, W.K. Chan, D.L. Hart, K. Luther, J.M. Tarascon: Science **252**, 944 (1991)
4.12 H.L. Kao, J. Kwo, R.M. Fleming, M. Hong, J.P. Mannaerts: Appl. Phys. Lett. **59**, 2748 (1991)
4.13 A. Kumar, L. Ganapathi, J. Narayan: Appl. Phys. Lett. **56**, 2034 (1990)
4.14 S.H. Liou, K.D. Aylesworth, N.J. Ionno, B. Johs, D. Thopson, D. Meyer, J.A. Woollam, C. Barry: Appl. Phys. Lett. **54**, 760 (1989)
    B. Johs, D. Thompson, N.J. Ionno, J.A. Wollam, S.H. Liou, A.M. Herman, Z.Z. Sheng, W. Kiehl, Q. Shams, X. Fei, L. Sheng, Y.H. Liu: Appl. Phys. Lett. **54**, 1810 (1989)
4.15 G. Koren, A. Gupta, E.A. Giess, A. Segmuller, R.B. Laibowitz: Appl. Phys. Lett. **54**, 1054 (1989)
4.16 R.A. Schweinfurth, C.E. Platt, M.R. Teepe, D.J. Van Harlingen: Appl. Phys. Lett. **61**, 480 (1992)
4.17 J.S. Horwitz, D.B. Chrisey, K.S. Grabowski, R.E. Leuchtner: Surf. Coating Tech. (in press)

4.18 A.T. Findikoglu, S. Bhattacharya, C. Doughty, M.S. Pambianchi, Q. Li, X.X. Xi, S.M. Anlage, R.E. Fahey, A.J. Strauss, J.M. Phillips and T. Venkatesan: Appl. Phys. Lett. (in press)
4.19 X.D. Wu, R.E. Muenchausen, N.S. Nogar, A. Pique, R. Edwards, B. Wilkens, T.S. Ravi, D.M. Hwang, C.Y. Chen: Appl. Phys. Lett. **58**, 304 (1991)
4.20 S.M. Garrison, N. Newman, B.F. Cole, K. Char, R.W. Barton: Appl. Phys. Lett. **58**, 2168 (1991)
4.21 X.D. Wu, R.C. Dye, R.E. Muenchausen, S.R. Foltyn, M. Maley, A.D. Rollet, A.R. Garcia, N.S. Nogar: Appl. Phys. Lett. **58**, 2165 (1991)
4.22 C.B. Eom, R.J. Cava, R.M. Fleming, J.M. Phillips, R.B. van Dover, J.H. Marshall, J.W.P. Hsu, J.J. Krajewski, W.F. Peck, Jr. Science, **258**, 1799 (1992)
4.23 K.M. Satyalakshmi, R.M. Mallya, K.V. Ramanathan, X.D. Wu, B. Brainard, D.C. Gautier, N.Y. Vasanthacharya, M.S. Hegde: Appl. Phys. Lett. **62**, 1233 (1993)
4.24 S.R. Foltyn, R.C. Dye, K.C. Ott, K.M. Hubbard, W. Hutchinson, R.E. Muenchausen, R.C. Estler, X.D. Wu: Appl. Phys. lett. **59**, 594 (1991)
4.25 D. Lowndes: Oak Ridge National Labs., private communication
4.26 This work was performed at Neocera, Inc. (College park, MD) as part of a DOE SBIR Phase I Program.
T. Venkatesan, X.D. Wu, R. Muenchausen, A. Pique: MRS Bulletin No. XVII 54 (1992)
4.27 J.C.S. Kools, C.J.C.M Niellesen, S.H Brongersma, van de Riet E, and J. Dieleman: J. Vac. Sci. Technol. A **10**, 1809 (1992)
4.28 G. Koren, R.J. Baseman, A. Gupta, M.I. Lutwyche, R.B. Laibowitz: Appl. Phys. Lett. **56**, 2144 (1990)
4.29 C.E. Otis, R.W. Dreyfus: Phys. Rev. Lett. **67**, 2102 (1991)
4.30 K.M. Yoo, R.R. Alfano, X. Guo, M.P. Sarachik, L.L. Issacs: Appl. Phys. Lett. **54**, 1278 (1989)
4.31 Q.Y. Ying, D.T. Shaw, H.S. Kwok: Appl. Phys. Lett. **53**, 1762 (1988)
4.32 T. Venkatesan, X.D. Wu, A. Inam, Y. Jeon, M. Croft, E.W. Chase, C.C. Chang, J.B. Wachtman, R.W. Odom, F. Radicati di Brozolo, C.A. Magee: Appl. Phys. Lett. **53**, 1431 (1988)
4.33 D.B. Geohegan, D.N. Mashburn: Appl. Phys. Lett. **55**, 2766 (1989)
4.34 L. Lynds, B.R. Weinberger, G.G. Peterson, H.A. Krasinski: Appl. Phys. Lett. **52**, 320 (1988)
L. Lynds, B.R. Weinberger, D.M. Potrepka: *Photochemistry in Thin Films*. SPIE **1056**, 249 (1989)
4.35 R. Kelly, R.W. Dreyfus: Nucl. Instrum. Methods Phys. Res. B. **32**, 341 (1988)
R. Kelly, R.W. Dreyfus: Surf. Sci. **198**, 263 (1988)
4.36 A. Gupta, B. Braren, K.G. Kasey, B.W. Hussey, R. Kelly: Appl. Phys. Lett. **59**, 1302 (1991)
4.37 T. Okada, N. Shibamaru, Y. Nakayama, M. Maeda: Appl. Phys. Lett. **60**, 941 (1992)
4.38 T.J. Geyer, W.A. Weimer: Appl. Phys. Lett. **54**, 469 (1989)
4.39 C. Girault, D. Damiani, J. Aubreton, A. Catherinot: Appl. Phys. Lett. **54**, 2035 (1989)
4.40 X.D. Wu, B. Dutta, M.S. Hegde, A. Inam, T. Venkatesan, E.W. Chase, C.C. Chang, R. Howard: Appl. Phys. Lett. **54**, 179 (1989)
4.41 J.P. Zheng, Q.Y. Ying, S. Whitanachi, Z. Q. Huang, D.T. Shaw, H.S. Kwok: Appl. Phys. Lett. **54**, 954 (1989)
4.42 C. Girault, D. Damiani, C. Champeaux, P. Marchet, J.P. Mercurio, J. Aubreton, A. Catherinot: Appl. Phys. Lett. **56**, 1472 (1990)
4.43 M. Kanai, T. Kawai, S. Kawai: Appl. Phys. Lett. **58**, 771 (1991)
4.44 N. Sughi, K. Kubo, M. Ichikawa, K. Yamamoto, H. Yamaguchi, S. Tanaka: Jpn. J. Appl. Phys. **31**, L1024 (1992)
4.45 A. Gupta, B.W. Hussey: Appl. Phys. Lett. **58**, 1211 (1992)
4.46 M.Y. Chern, A. Gupta, B.W. Hussey: Appl. Phys. Lett. **60**, 3045 (1992)
4.47 Q.Y. Ying, H.S. Kwok: Appl. Phys. Lett. **56**, 1198 (1990)
4.48 C.T. Rogers, A. Inam, M.S. Hegde, B. Dutta, X.D. Wu, T. Venkatesan: Appl. Phys. Lett. **55**, 2032 (1989)
J.B. Barner, C.T. Rogers, A. Inam, R. Ramesh S. Bersey: Appl. Phys. Lett. **59**, 742 (1991)
4.49 T. Hashimoto, M. Sagoi, Y. Mizutani, J. Yoshida, K. Mizushima: Appl. Phys. Lett. **60**, 1756 (1992)
4.50 F.C. Wellstood, J.J. Kingston, J. Clarke: Appl. Phys. Lett. **57**, 1930 (1991)

4.51 Q. Li, X.X. Xi, X.D. Wu, A. Inam, S. Vadlamannati, W.L. McLean, T. Venkatesan, R. Ramesh, D.M. Hwang, J.A. Martinez, L. Nazar: Phys. Rev. Lett. **64**, 3086 (1990)
4.52 D.H. Lowndes, D.P. Norton, J.D. Budai: Phys. Rev. Lett. **65**, 1160 (1990)
4.53 D.P. Norton: Phys. Rev. Lett. **66**, 1537 (1991)
4.54 R.F. Wood: Phys. Rev. Lett. **66**, 829 (1991)
4.55 J. Pennycook: Phys. Rev. Lett. **67**, 765 (1991)
4.56 X.Xi, Q. Li, C. Doughty, C. Kwon, S. Bhattacharya, A.T. Findikoglu, T. Venkatesan: Appl. Phys. Lett. **59**, 3470 (1991)
4.57 X.Xi, T. Venkatesan, Q. Li, C. Doughty, A. Walkenhorst, C. Kwon: Appl. Phys. Lett. **61**, 2353 (1992)
4.58 A. Walkenhorst, C. Doughty, X.X. Xi, Q. Li, C.J. Lobb, S.N. Mao, T. Venkatesan: Phys. Rev. Lett. **69**, 2709 (1992)
4.59 P. Schwab, X.Z. Wang, D. Bäuerle: Appl. Phys. Lett. **60**, 2023 (1992)
4.60 K.S. Harshavardhan, R. Ramesh, T.S. Ravi, S. Sampere, A. Inam, C.C. Chang, G. Hull, M. Rajeswari, T. Sands, T. Venkatesan, M. Reeves: Appl. Phys. Lett. **59**, 1638 (1991)
4.61 Y. Iijima, N. Tanabe, O. Kohno, Y. Ikeno: Appl. Phys. Lett. **60**, 769 (1992)
4.62 X.D. Wu, R.E. Meunchausen, S. Foltyn, R.C. Estler, C. Flamme, N. Nogar, A.R. Garcia, J. Martin, J. Tesmer: Appl. Phys. Lett. **56**, 1481 (1990)
4.63 X.D. Wu, R.E. Muenchausen, S, Foltyn, R.C. Estler, R.C. Dye, A.R. Garcia, N. Nogar, P. England, R. Ramesh, D.M. Hwang, T.S. Ravi, C.C. Chang, T. Venkatesan, X.X. Xi, Q. Li, A. Inam: Appl. Phys. Lett. **57**, 523 (1990)
4.64 J.A. Greer, J. B. Hook: SPIE **1377**, 79 (1990)
4.65 R.E. Muenchausen, R.C. Dye, X.D. Wu, L. Luo, D.W. Cooke: Appl. Phys. Lett. **59**, 1374 (1991)
4.66 M.F. Davis, J. Wosik, K. Forster, S.C. Deshmukh, H.R. Rampersad, S. Shah, P. Siemsen, J.C. Wolfe, D.J. Economu: J. Appl. Phys. **69**, 7182 (1991)
4.67 S. Witanachchi, H.S. Kwok, X.W. Wang, D.T. Shaw: Appl. Phys. Lett. **53**, 234 (1988)
4.68 R.K. Singh, J. Narayan, A.K. Singh, J. Krishnaswamy: **54**, 2271 (1989)
4.69 For example, for a US $20,000 difference the laser pulse energy can be increased by 50%, while for a similar price difference the repetition rate can be increased by an order of magnitude (private communication with G. Zaal of Lambda Physik)
4.70 I.W. Boyd, E. Fogarassy, M. Stuke (eds.): *Surface Processing and Laser Assisted Chemistry*, Proc. of the 1990 E-MRS Spring Conf, Strasbourg, France (1990) (North-Holland, Amsterdam 1991)

5. Interaction of Laser Radiation with Organic Polymers

R. Srinivasan

With 21 Figures

Organic polymers form a class of organic molecules of considerable technological importance in the field of materials processing. For this reason, their interactions with laser radiation have attracted attention from diverse groups of scientists who have studied these interactions from perspectives that have ranged from applied physics to surgery. Since the general field of ultraviolet-laser interactions with organic polymers has been reviewed more than once in recent years, it is the intention of this review to concentrate on developments that have taken place over the last five years and to include infrared-laser interactions with polymers in order to compare results which were obtained using different wavelengths of the laser.

5.1 History

Organic molecules interact with photons in ways that are unlike the pathways by which metals and inorganic compounds react with photons. Therefore, the subject of organic photochemistry occupies a niche that is unique. UltraViolet (UV) photons when absorbed by organic molecules give rise to electronic excitation in the first instance whereas InfraRed (IR) photons lead to vibrational and rotational excitation. The reactions that the electronically excited organic molecules subsequently undergo are so varied and so dependant upon the chemical structure of the molecule, the wavelength of the photon and the medium in which the reaction is carried out that the study of organic photochemistry has evolved over the past 100 years into a vast field [5.1]. In contrast, excitation by IR photons was not observed to lead to interesting effects until the advent of the laser. The interaction of IR-laser radiation with organic molecules gives rise to multiphoton excitation over the vibrational manifolds of ground electronic states which is then followed by thermal decomposition. The interaction of UV-laser pulses with small polyatomic molecules has received considerable attention following the early reports [5.2] that in these molecules multiphoton excitation to upper electronic states results in ionization and decomposition by a variety of pathways [5.3].

It was first reported in 1982 [5.4, 5] that when pulsed UV-laser radiation falls on the surface of an organic polymer, the material at the surface is spontaneously etched away to a depth of 0.1 μm to several microns. The principal features of this phenomenon which readily distinguished the interactions of UV-laser pulses from visible or IR-laser pulses were the control that can be exercised over the depth of the etching by controlling the number of pulses and the fluence of the laser and the lack of detectable thermal damage to the substrate. The result is an etch pattern in the solid with a geometry that is defined by the light beam. Within a period of a year after the first report, other groups confirmed these observations in several other polymers and at different ultraviolet wavelengths [5.6–9]. The possibility of extending the process to biological tissue was reported in 1983 [5.10].

Research devoted to understanding the science and developing the technology behind UV-laser ablation of polymers has grown at a surprising rate over the past decade. The subject has been reviewed most recently in 1989 [5.11] and in 1992 [5.12]. The focus of the former is the science behind the process while the latter concentrates more on the applications.

The activity in the field of laser interactions with organic polymers is not limited to the use of UV radiation. In fact, many articles have been published concurrently which aim to compare the results from UV-and IR-laser radiation. The aim in most of these studies is to gain an understanding of the detailed mechanism of the laser interactions at various photon wavelengths. It would not be an exaggeration to say that the interactions of IR lasers with organic polymers which were investigated cursorily in the first decade (1960–1970) after the invention of the laser have been investigated in a more sophisticated way this second time around. The focus of this review will be the interactions of nanosecond, ultraviolet-laser pulses from an excimer laser with the few organic polymers on which a variety of data has been obtained but the results of studies that have been made under other conditions of pulse width and wavelength will be cited where appropriate.

5.2 Characteristics of UV-Laser Ablation

Nearly all organic polymers show moderate to intense absorption in the ultraviolet region. These absorptions are usually ascribed to electronic transitions from a ground singlet to the first excited singlet states. The unique features in the UV-laser ablation of polymers are encountered only in those wavelength regions in which such electronic absorptions exist.

The ablation of the surface of a polymer by a UV-laser pulse is a function of the energy deposited in the solid in unit time. If a typical UV pulse has a Full Width at Half Maximum (FWHM) of 20 ns, an energy of 450 mJ and the size of the beam at the polymer surface is 1.5 cm^2, the fluence at the surface will be 1.5 × 10^7 W/cm^2. When this pulse strikes the surface a loud audible report will be heard and, depending upon the wavelength, 0.01–0.1 µm of the material would have been etched away with a geometry that is defined by the light beam. If this experiment is performed in air, a bright plume will be ejected from the surface

5. Interaction of Laser Radiation with Organic Polymers   109

and will extend to a few millimeters. Typically, UV-laser ablation is carried out with a succession of pulses. The etching of the surface is a linear function of the number of pulses when the polymer is a strong absorber at the laser wavelength. In the case of a weak absorber such as PolyMethyl MethAcrylate (PMMA) at 248 nm, a phenomenon which has been termed "incubation" [5.13] is observed. It results in the first few pulses not causing any etching at all as shown in Fig. 5.1 [5.14] and even giving rise to a raised surface. But after a few pulses – the exact number depending upon the fluence, the wavelength and the absorption of the polymer – the system settles down to a constant etch depth per pulse. Research devoted to the establishment of the cause of this "incubation" effect [5.14–16] suggests that more than one reason may be important. The value for the etch depth per pulse is usually averaged over hundreds of pulses in order to minimize the uncertainties that are introduced in the measurement of the etch depth as well as incubation effects where they exist. Average values for the etch depth/pulse are reproducible to within the uncertainties in the measurement of the fluence provided the absorption characteristics of the polymer are well-controlled [5.17].

The use of UV-laser ablation to etch polymer surfaces has become an established technology [5.18]. The potential for such use was an important factor in the progress that has been made in the investigation of this phenomenon [5.12]. With this perspective in mind, nearly all of the etch data have been presented in the literature as a plot of etch depth/pulse versus log fluence. It is

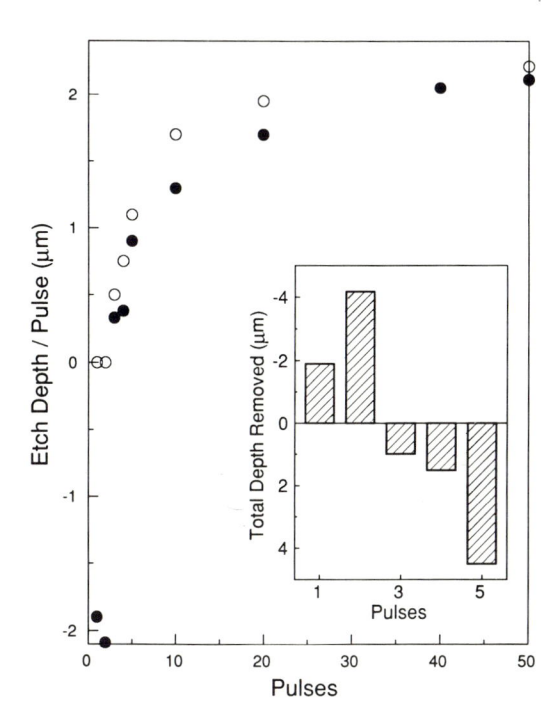

Fig. 5.1. Average etch depth/pulse for the ablation of PMMA by 248 nm laser pulses (fluence 2.1 J/cm^2). The filled and open data points show the values before and after washing with a solvent. Inset: Actual depth of etched surface during the first five pulses. The first two pulses lead to an elevation of the surface [5.14]

intuitive that the interaction of photons with the polymer should involve the power density (power/unit area) rather than the fluence. Discussions based on the fluence are acceptable only so long as the pulse width is constant. The major portion of the data that were published until the last few years involved the use of commercial excimer lasers whose pulse widths fall within the narrow range of 15–35 ns (FWHM). Therefore, the use of fluence rather than the power density as the laser parameter that controlled the etch depth (at a given wavelength and for a particular polymer) was acceptable. In the last three years, there has been considerable work done on the interactions of femtosecond UV-laser pulses as well as microsecond UV-laser pulses on organic polymers. While these results (which will be discussed later) are still not definitive, they promise to add new insights to this field.

Three examples of the "standard" etch-depth/pulse versus log-fluence plots are shown in Figs. 5.2 [5.19] and 5.3 [5.20]. Figure 5.2 represents a system which is a strong absorber at the laser wavelength, i.e., the polyimide (Kapton™) at 248 nm. Figure 5.3 shows two curves which are those of a moderate absorber (PMMA at 193 nm) and a weak absorber (PMMA at 248 nm). A detailed

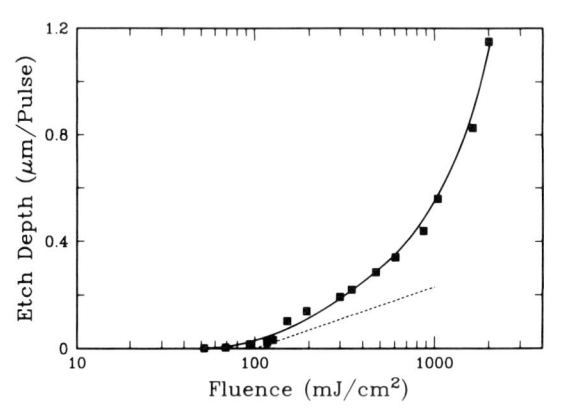

Fig. 5.2. Plot of etch depth/pulse vs log fluence for etching of a Kapton™ film at a laser wavelength of 308 nm. Error limits are shown in the value at 0.9 J/cm². The solid line is meant for visualization, the broken line represents the slope calculated from the reciprocal of the ultraviolet absorptivity [5.19]

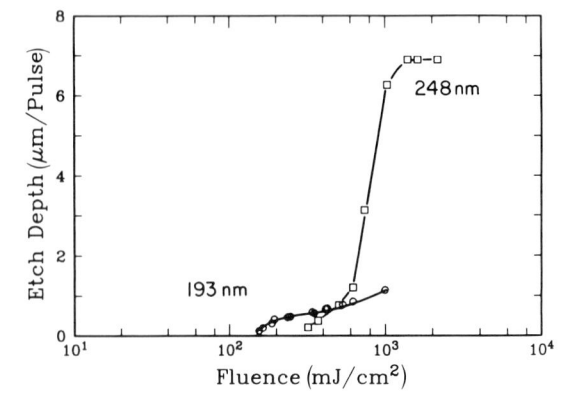

Fig. 5.3. Plot of etch depth vs log fluence in the etching by laser ablation of PMMA at 193 nm and 248 nm [5.20]

examination of these plots and the thought process behind them is appropriate at this point since it has been the basis for the substantial literature that has grown (nearly 25 publications) which attempts to interpret or model these curves. Historically, the etch-depth/pulse versus log-fluence plot originated in the work of two independent groups [5.6, 9] who simultaneously showed that the linear region of this plot can be fitted to the simple equation

$$l_f \text{ (etch depth/pulse)} = \alpha^{-1} \log (F/F_0), \tag{5.1}$$

where F_0 is the threshold fluence at which etching is first observed, α is the absorptivity of the polymer at the laser wavelength and F is a fluence above the threshold value at which ablation and etching are conducted. As data over a wide range of fluences and on a number of polymers were reported, it became apparent that the linear region of the etching prevailed only over a short range of fluence, as seen in all three of the plots in Figs. 5.2 and 3. The fluence threshold which was easy to define in a linear plot was shown to be ill-defined as the etch curve approached the abscissa assymptotically. Even if a linear region is picked out from the etch curve, its slope has never been observed to be $1/\alpha$, where α is the absorptivity as defined by Beer's Law. This has given rise to the pragmatic suggestion that the absorptivity can be actually measured from the slope to give an "effective absorption coefficient" [5.21]. The very first laser pulse that strikes a polymer surface sees a virgin material. At the end of this pulse, the material is etched to a depth that is invariably less than the depth to which the photons have penetrated the material. If the material in the penetration depth minus the etch depth is transformed by the photons without causing them to ablate, the photons in the next pulse would see a surface whose absorption properties can be significantly different. Analysis of the exposed surface by X-ray photoelectron spectroscopy [5.22–26] as well as by infrared spectroscopy [5.27] has confirmed the existence of such a transformed surface layer. What has not been established is the depth to which such a transformed layer can exist and how its absorption properties differ from those of the virgin material. In any case, when the system settles down to a constant etch depth/pulse at a given wavelength and fluence, each pulse of photons must be interacting with the surface in a consistent manner. The time in which this interaction takes place in relation to the pulse width has been a subject of intense research activity as mentioned already.

It is useful to replot the data in Figs. 5.2 and 3 (248 nm) which represent the extreme examples of a strong absorber and a weak absorber in two other ways in order to bring out other relationships between these two quantities. In the first of these (not shown) the data are plotted as etch depth/pulse versus fluence (instead of log fluence). This merely serves to lessen the importance of the data points near the threshold fluence and increases the importance of the data that were obtained at fluences > 1 J/cm^2. There has been a recent trend to extend the etch data at a constant pulse width [5.28] and at different pulse widths [5.29] to fluences as large as 50 J/cm^2. In Figs. 5.4a, b, the data are plotted as log (etch depth/pulse) versus $1/F$. The reasoning behind this plot is as follows: the etch depth/pulse (l_f), when multiplied by the cross-section of the exposed area

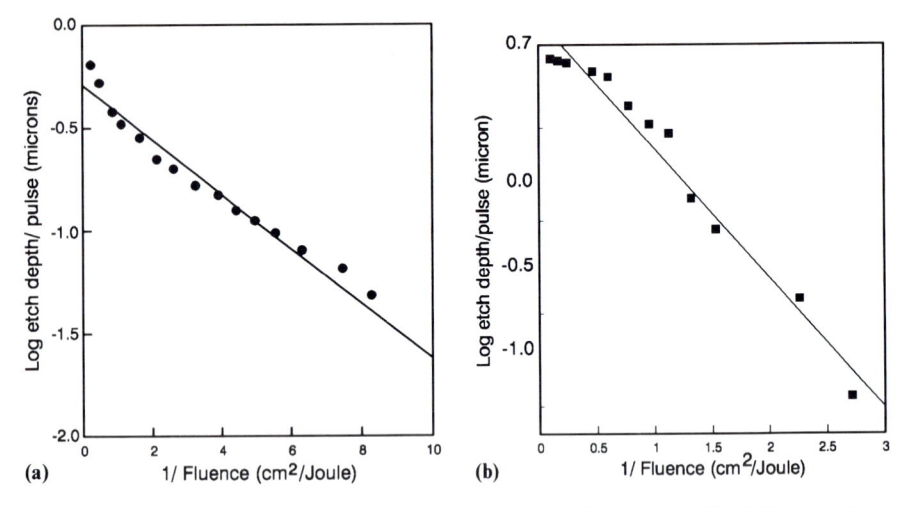

Fig. 5.4. Plots of log (etch depth/pulse) vs. 1/fluence. (**a**) polyimide at 308 nm; (**b**) PMMA at 248 nm

(1 cm²), the density (ρ), and divided by M, the molecular weight of the monomer unit gives the amount of monomer moles that are etched in the time of one pulse. If the actual duration of the pulse is τ, then

$$l_f \rho / M \tau = \text{rate of decomposition of the polymer}. \tag{5.2}$$

Since the etching reaction should be of zero order in the concentration of the monomer units, the rate of decomposition is simply equal to k, the rate constant for the process under the reaction conditions. The fluence F at the polymer surface measures the energy supplied to the polymer volume that undergoes ablation. If all of it or a constant fraction of it. (depending upon the model used to interpret the mechanism of the decomposition and ablation of the polymer) raises the temperature of the polymer volume that undergoes ablation, then

$$F \propto \Delta T, \tag{5.3}$$

where ΔT is the temperature rise and the proportionality constant would involve the reflectivity of the polymer surface, its absorptivity, the specific heat and the density of the polymer [5.12]. At the threshold for ablation, ΔT can be as high as 1000°C [5.12]. If the polymer surface is assumed to be at room temperature or 25°C before the laser pulse strikes the surface, its final temperature would be sufficiently high so that $\Delta T \approx T$ be set, where T is the final temperature on the absolute scale. The approximation would be poor at the threshold but would not introduce a significant error at fluences ≥ 0.5 J/cm². Thus a plot of log (etch depth/pulse) versus F^{-1} should be equivalent to an Arrhenius plot and should be linear. Its slope should be a measure of the activation energy of the process involved in the ablation reaction. A further discussion of these plots will be found in a later section. It is sufficient to point out here that the data plotted in this manner in Figs. 5.4a, b are reasonable straight lines which indicate that a single activated process prevails under each of these ablation conditions.

5.3 Chemical Physics of the Ablation Process

A pictorial representation of the interaction of a laser pulse with a polymer surface is shown in Fig. 5.5. At the top of the figure, a stream of photons from a single laser pulse is shown to fall on a polymer surface. It is absorbed in a depth which is believed to be of the order of 10–30 µm for a weak absorber, 1–5 µm for a moderate absorber and < 1 µm for a strong absorber. While the absorption coefficient that applies to this absorption process is in dispute (*vide supra*), it is accepted that the penetration of photons into the material will fall off as a logarithmic function. The middle part of the figure shows that the absorption of

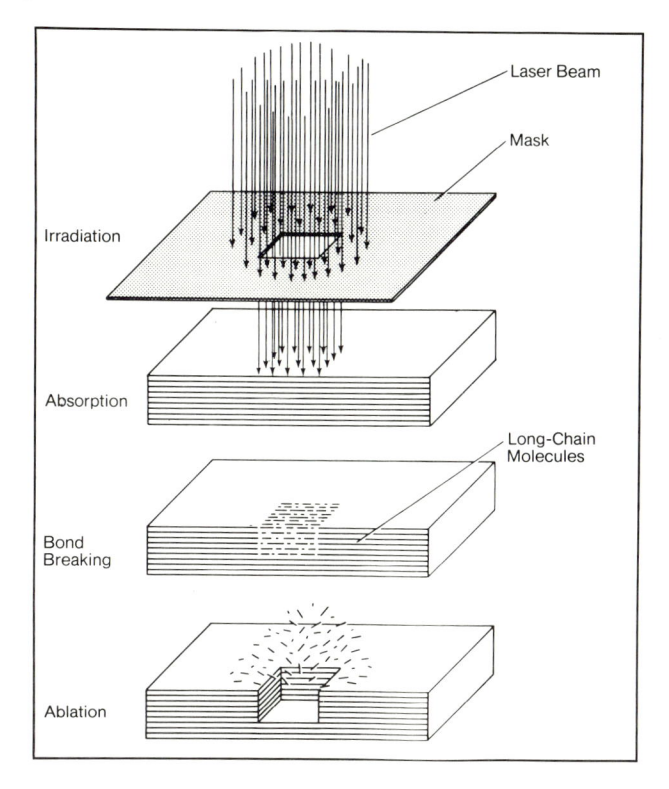

Fig. 5.5. Hypothetical steps in the interaction of a UV-laser pulse with a polymer. Top: The laser radiation which is defined by a mask is absorbed; Middle: Chemical bonds in the polymer are broken by the photon energy; Bottom: The products ablate at supersonic velocities leaving an etched sample

the photons causes the break-up of the polymer chains. The exact mechanism by which this happens is still under vigorous debate. The bottom of the figure shows the ablation step in which the products of the decomposition of the polymer are being ejected at supersonic velocities. The detailed nature of the latter two steps has attracted the most work in the past few years. Before reviewing these, it will be useful to summarize the earlier work that has already been the subject of two reviews [5.11, 12].

5.3.1 Ablation Products

A knowledge of the composition of the material that is ablated from the surface of a polymer is of importance in understanding the chemistry of the process. Chemical analysis of the ablation products has not received as much attention as physical aspects of ablation. This can be attributed to the wide range of products such as atoms, diatomics, small molecules and solid fragments of the polymer that are ejected during ablation. No single analytical technique is capable of determining all of these species. There is also evidence that the composition of the products changes with the wavelength used (even within the ultraviolet spectrum [5.30, 31] and certainly in going from ultraviolet to infrared wavelengths), the repetition rate of the pulses (since this will affect the dissipation of heat energy that is left in the substrate from one pulse to the next), the chronology of the pulses and the absolute value of the fluence. The products may also react with the oxygen and the nitrogen in air [5.32] even though the etch depth/pulse is not sensitive to the pressure of air at the polymer surface. The net quantity of products that is formed is of the order of only milligrams in typical experiments on a laboratory scale. This, when combined with the complexity of the products, increases the difficulty in the analysis.

The work that was reported prior to 1989 has been reviewed in considerable detail [5.11]. In subsequent work, the most important discovery has been in the ablation of the polyimide (Kapton™), which is formed by the condensation of pyromellitic dianhydride and oxy-dianiline. The formula of this material is shown in Fig. 5.6. It is extraordinarily stable to heating in air up to 400°C and is not attacked by any of the common solvents [5.33]. Its susceptibility to UV-laser photons can be attributed to the four carbonyl groups which are probably the first to be eliminated as CO (Fig. 5.7). The nitrogen is detected as CN [5.19] and ultimately leads to HCN. The rest of the molecule is so rich in carbon and so poor in hydrogen content that it is expected to break up exclusively to small fragments such as C, C_2, C_3, CH and C_2H_2. What is remarkable is that within the hemispherical blast wave that spreads from the polymer surface in air, these

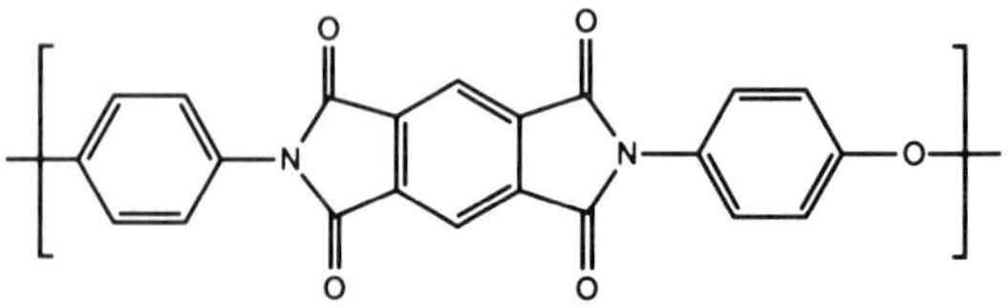

Fig. 5.6. Formula of the polimide, (Kapton™)

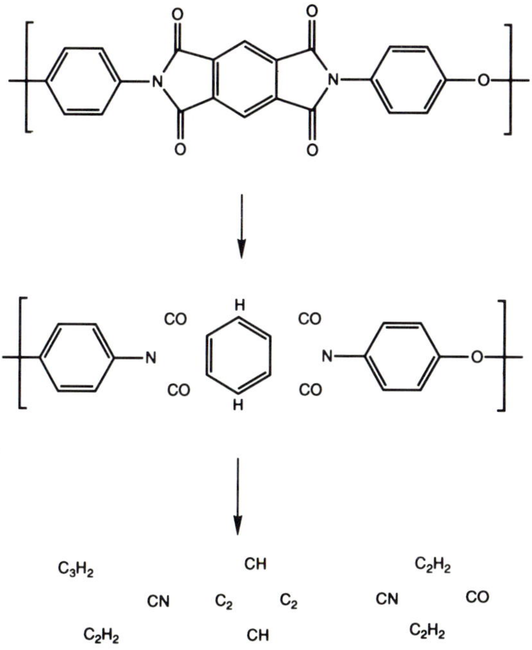

Fig. 5.7. A probable mechanism for the decomposition of the polyimide (Kapton™) on the absorption of UV photons

fragments seem to polymerize to give carbon clusters up to 1600 amu [5.34–37]. Elemental carbon which is deposited as a black soot on the surrounding surface can be looked upon as the final product of this polymerization process. The control of the formation of this carbon debris has attracted considerable work [5.38] because it is a detriment to the use of UV-laser ablation as a technology in the semiconductor processing industry. An atmosphere of pure oxygen leads to partial oxidation of the carbon [5.32] and an atmosphere of hydrogen or helium seems to eliminate the debris totally [5.38]. In an industrial environment, these solutions have not been acceptable. The soot is still removed only by vigorous blowing on the surface at a glancing angle and suction from the opposite side of the ablated area.

5.3.2 Time Profile of Ablation

A knowledge of the timing of the ablation process is fundamental to an understanding of the chemical physics of the phenomenon. The time-dependent probing methods that have been used to date include acoustic methods [5.39–47], fast photography of the ablating surface and the plume that rises from it by using a conventional camera [5.48–52], a streak camera [5.53], or by Schlieren photography [5.54], time-resolved reflectivity [5.55–57] and beam deflection measurements [5.46, 58, 59], absorption spectroscopy [5.41, 60–63] and emission spectroscopy [5.64–66]. From a historical perspective, even the very first measurements in 1984 and 1985 showed two of the serious complicating factors that have made these measurements difficult to interpret. The emission at various distances from the polymer surface was timed and it was concluded [5.63, 64] that in the etching of the polyimide films by 193 nm pulses (15 ns FWHM), the emission from the plume had a fast component that appeared simultaneously with the laser excitation, taking into account the ~ 30 ns response time of the photomultiplier that was used in the detection, but a slower component lasted 10–100-fold longer than the laser pulse itself. The emission from the plume in the ablation of PMMA at 193 nm (20 ns FWHM) was studied spectroscopically and the peak in the intensity of the emission of CH radicals at various distances from the surface was timed in order to calibrate its velocity [5.66]. These data also placed the beginning of the emission signal at times of the order of the laser pulse. It was pointed out that "it is possible that the photodissociation processes responsible for creating the emission in the plume are separate and subsequent to the breaking of the polymeric bonds which cause ablation". Both these studies showed that the UV-laser ablation process is complex and that the study of one variable as a function of time cannot, in itself, give unique answers to the time profile of the total process.

Some of the problems in conducting these experiments and the conflicts in the interpretation of the data can be illustrated by analyzing the published results on two representative polymers: polyimide (Kapton™) and PMMA.

a) Polyimide

In 1986, the time profile of the ablation process in this polymer was measured [5.39] by recording the electrical signal from a wide-bandwidth Poly(VinyliDeneFluoride)(PVDF) piezoelectric transducer that was attached to the polyimide film generated when the latter was irradiated with a UV-laser pulse. Typical data at two wavelengths are shown in Fig. 5.8. A stress wave was observed to begin with a delay of ~ 4–6 ns after the start of the laser pulse. For the 16 ns (FWHM) 193 nm pulse, the duration of the ablation (FWHM), as indicated by the stress pulse, was 16–20 ns, whereas for the longer and more structured 308 nm pulse ($\simeq 24$ ns) the duration was 50–70 ns but decreased to ≤ 25 ns as the fluence increased above 200 mJ/cm^2. These compressive stress pulses persisted down to fluences well below the "threshold" fluence for etching that was determined from etch-depth measurements. The precision of these photoacoustic experiments depends upon the minimization of the rise time of the acoustic detector relative to the width of the laser pulse. For the same reason, the thickness of the polymer sample should also be small in order to keep the transit time of the stress wave to a minimum. The origin of the stress wave is most probably due to the enormous pressure that is generated in the polymer film by the decomposition of some of the chemical bonds in the polymer network to create many small molecules in the same volume.

In the past few years, the acoustic technique has been developed considerably in order to provide a sensitive method for the detection of the

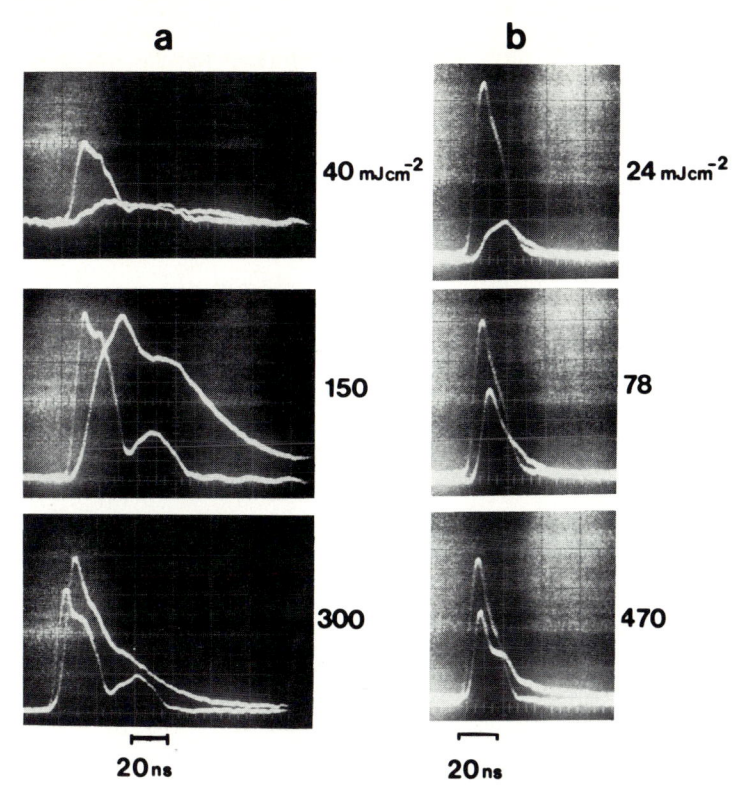

Fig. 5.8. Time-synchronized laser and stress pulses for polyimide film irradiated at various fluences using (a) 308 nm and (b) 193 nm radiation [5.39]

threshold for ablation [5.40–43, 47,59], the magnitude of the shock wave in confined systems [5.44], monitoring the etching of multilayers [5.47] and the detection of the end point for etching [5.45].

A direct "look" at the polymer surface as it is undergoing UV-laser ablation was first provided by fast (< 1 ns) photography in 1989 [5.48]. In this work, the sub-nanosecond dye-laser beam that illuminated the surface for the camera to photograph was pumped by the same UV-laser that caused the ablation. By delaying the probe pulse for suitable intervals, it was shown that at various fluences of 308 nm laser radiation above the ablation threshold, the progressive darkening of the polymer surface with time was indicative of the etching process taking place *even during the pulse*. In a more elaborate version of this photographic approach, later workers [5.49–52] were able to discriminate different features of this "darkening" of the polymer surface. Their experimental arrangement is shown in Fig. 5.9. By the use of a second excimer laser to pump the dye laser and thereby produce short visible (596 nm) laser pulses, it was possible to extend the time over which the photographs could be taken. A series of photographs were taken of the surface of a polyimide sample after a single pulse

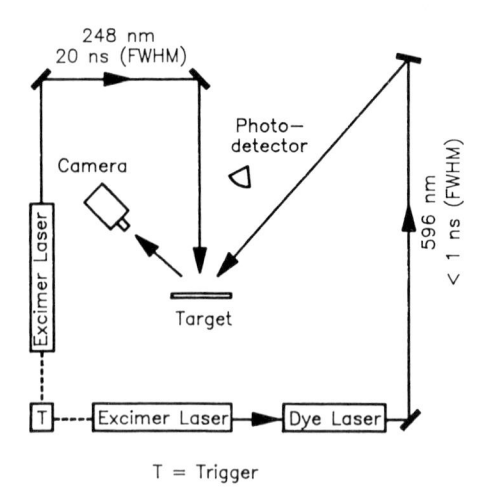

Fig. 5.9. Experimental arrangement to view the polymer surface during UV-laser ablation [5.49]

of 248 nm laser radiation had impinged on it. A fresh surface of the polymer was used for each exposure. Some of these results are shown in Fig. 5.10. The darkening of the surface is seen to be a transient feature which is due to a decrease in its reflectivity. Since the surface shows no sign that it is etched permanently in the sense that there is a depression in the irradiated area, the transient darkening is best attributed to a layer of gas bubbles that collect on the

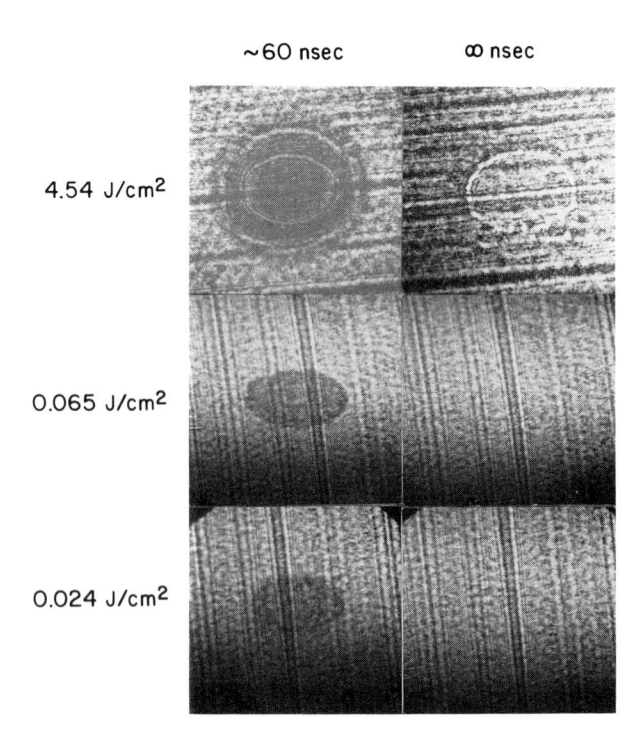

Fig. 5.10. Pictures of a polyimide surface ≈ 60 ns after the start of one single UV (248 nm; 20 ns FWHM) ablation pulse and ∞ time after the pulse [5.51]

surface above the irradiated volume and are ejected into the atmosphere (Fig. 5.11). The **primary** decomposition products of the UV-laser ablation of polyimide at this wavelength have been collectively seen to be gaseous from an experiment [5.67] in which a pyroelectric detector was used to sample the products. This irradiation was performed in a vacuum and the laser fluence was kept close to the threshold value for ablation. The detector was used simply to time the arrival of the products (Fig. 5.12). The products from polyimide (as from all strongly absorbing polymers) were observed to travel at supersonic velocities that averaged $> 1.5 \times 10^5$ cm s^1 even at the fluence threshold. The products were also seen to arrive at the detector in a compact burst that lasted from

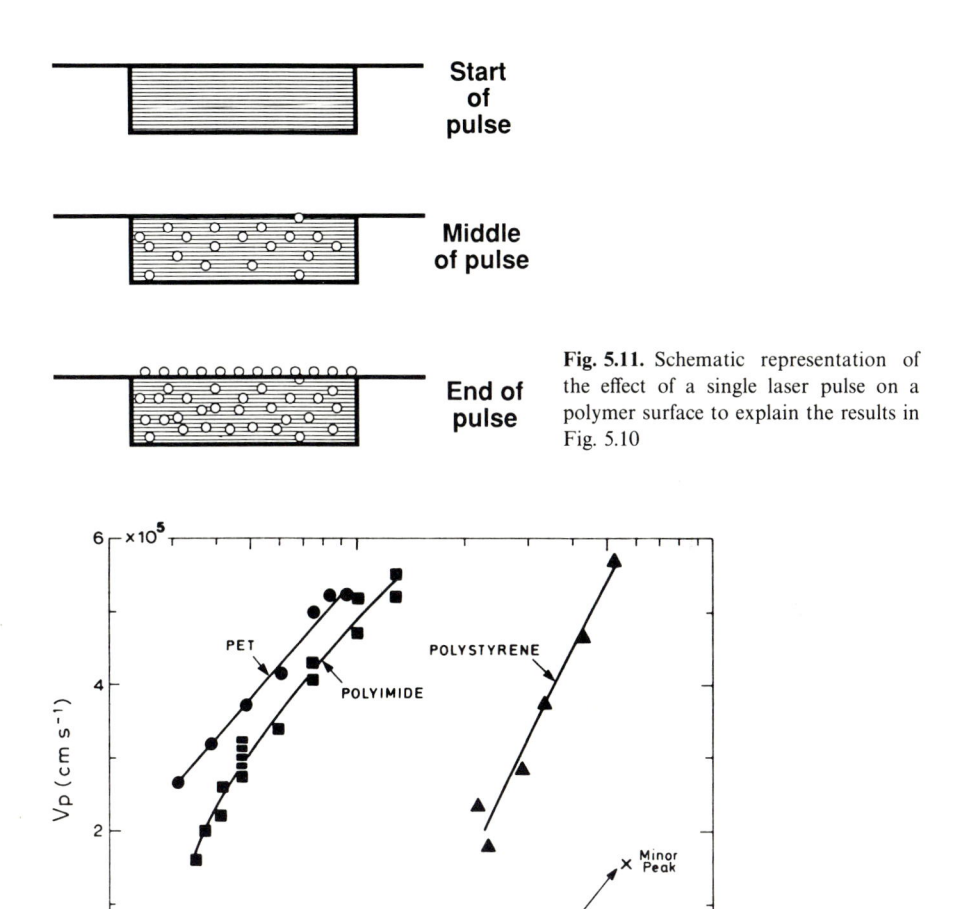

Fig. 5.11. Schematic representation of the effect of a single laser pulse on a polymer surface to explain the results in Fig. 5.10

Fig. 5.12. Plot of velocity of product fragments vs. fluences during ablation by UV (248 nm) laser pulses. The lines are meant to guide the eye [5.67]

10–60 μs (the start of the laser pulse was set as zero time) when the detector was 7.5 cm from the polymer surface. The photographs shown in Fig. 5.11 suggest that gas production can take place in the entire depth to which the photons penetrate but the polymer is sufficiently porous to these shallow depths to allow the gas to escape. Since no net etching of the surface is observed, this brings up the interesting question of: 'Does the fluence threshold really represent the threshold for etching or the threshold for explosive expulsion of the gaseous products'? The difference in the fluence between these two thresholds is probably not larger than a factor of two (0.02 J/cm^2 for gas production and 0.04 J/cm^2 for etching) but this points out the difficulty in relying on one set of measurements for establishing the nature of the interactions between the photons and the polymer surface. Quantitative studies [5.56, 58] of the time-resolved reflectivity of the surface of several polymers including polyimide during the laser pulse have recently been reported. Both the amplitude and the width (FWHM) of the reflected pulse were found to be sharply truncated when the laser fluence at 248 nm reached a value that was close to the ablation threshold that was established by acoustic measurements [5.40]. However, to model this in a quantitative way proved difficult since the cause of the truncation could not be readily established. The authors observe [5.56] that: "Many complex processes are taking place during laser ablation of polymers and it is difficult to relate them causally to the measured quantities and quantify their relative contribution".

More photographic studies were undertaken [5.50, 51] using the geometry shown in Fig. 5.13. In this arrangement, the gaseous products are seen to rise from the surface at supersonic velocities to form a hemispherical blast wave (Fig. 5.14). Since the probe-laser beam lights this bubble from behind with respect to the camera, dense material such as finely divided particles are clearly seen as black dots. Liquid droplets can be distinguished from solids and heavy refractive vapors from light gases.

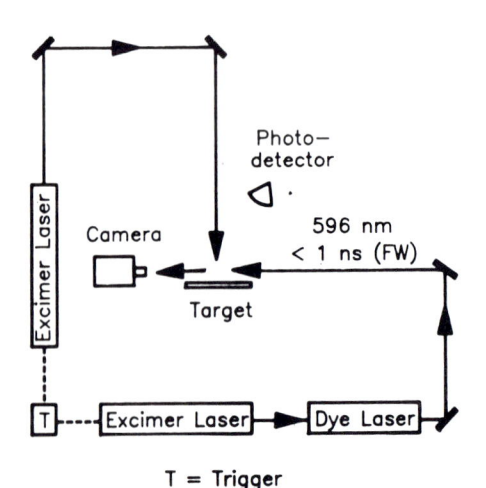

Fig. 5.13. Schematic representation of the arrangement to photograph the blast wave rising from a polymer surface during UV-laser ablation in an air medium [5.50]

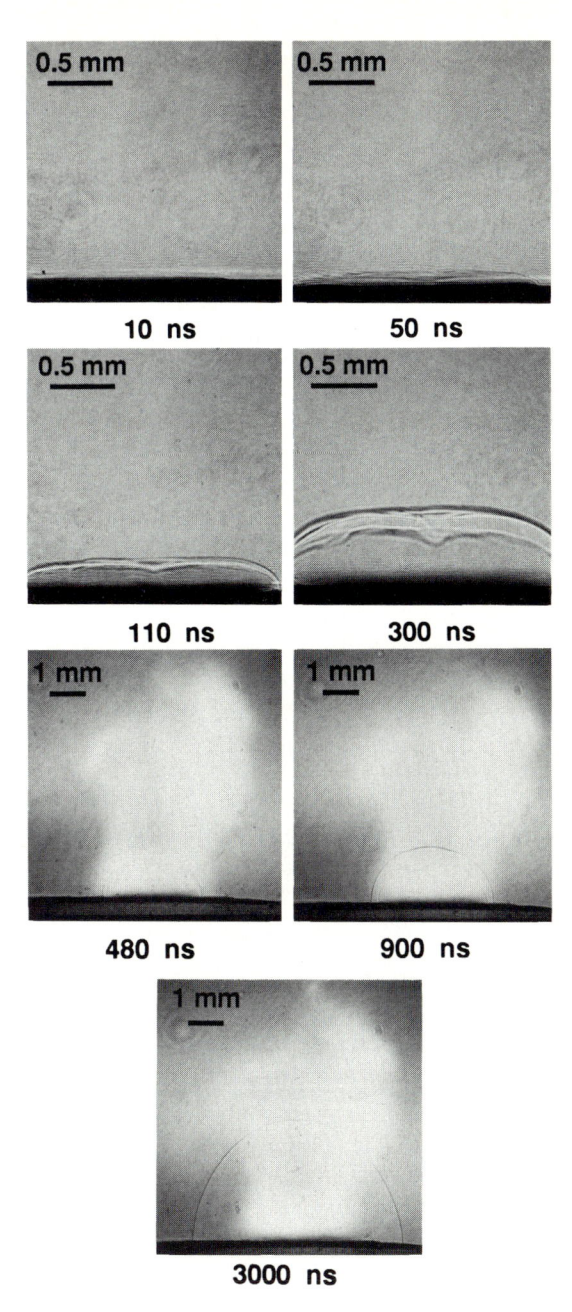

5. Interaction of Laser Radiation with Organic Polymers   121

Fig. 5.14. Successive photographs of the blast wave rising above a polyimide surface during ablation by a single pulse of 308 nm laser radiation. The contact front is visible at 300 ns. The luminous background after 480 ns is the radiant emission from the products [5.68]

The sequence of exposures in Fig. 5.14 shows the plume from the ablation of polyimide by a single 308 nm pulse. The first set of four shots at high magnification shows the emergence of a blast wave at 50 ns which is approximately the point at which > 90% of the laser photons had been delivered to the surface. The blast front and the contact front which follows it are clearly seen in the exposure at 300 ns. Subsequent exposures taken at 4 × smaller magnification show the growth of the blast wave both in the normal direction and parallel to the surface. While dense material and the fronts are "frozen" in the exposures by the visible-laser pulse, an emission that originates from the ablated material itself is not frozen in time. As a result, a constant bright background is seen in the later exposures. It is present in the first four frames as well except that it is not defined well because of the lack of contrast from the background lighting. The emission alone can be photographed by switching off the dye laser and leaving the camera shutter open for the duration of the ablation. It is more effective to use a streak camera [5.53] and photograph the development of the emission as a function of time. This is shown in Fig. 5.15.

It is generally accepted that the ablation of organic polymers by a UV-laser pulse is a volume explosion that is caused by a large rise in pressure and temperature in the volume of the material that undergoes ablation. The ejected gas compresses the material that is ahead of it and eventually forms a shock wave provided the thickness of the gas layer between the shock front and the driving gas layer is considerably greater than the mean free path of the ambient gas. The theory behind this as it applies to the present problem has been discussed elegantly [5.53]. At pressures greater than 15 mbar of ambient gas, it is derived that the shock conditions will be fulfilled at less than 250 µm from the surface of the polymer film. This is seen to be so in the exposure at 110 ns in Fig. 5.14. The motion of the shock front can be tracked from the photographs by measuring the position of the front as a function of time. The velocity of the shock wave has initial values that are supersonic (Fig. 5.16) and remain so at times as long as 10 µs. In this region, the motion of the front does not obey a $t^{0.4}$ relationship, where t is time [5.53], but instead has exponents that are 0.65 and larger. However, when the progress of the blast wave has been tracked to distances far greater than those shown in Fig. 5.14 (centimeters instead of < 1 cm) and times greater than 1 ms, the $t^{0.4}$ relationship has been found to hold [5.54]. An exponent that is > 0.4 would mean that the blast wave is travelling faster then predicted by the blast-wave theory. This could be the case

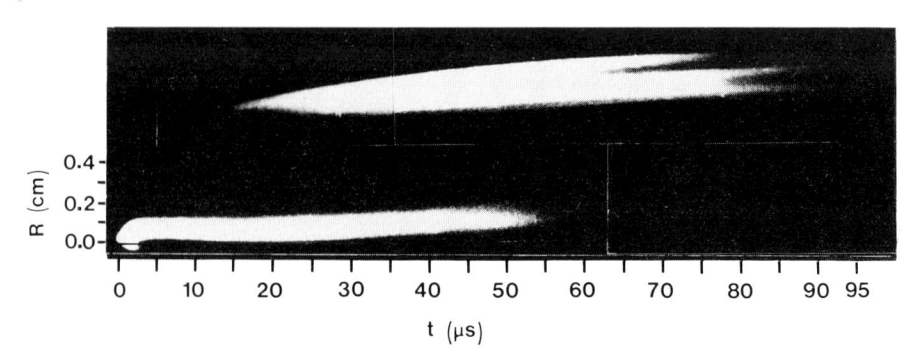

Fig. 5.15. Streak camera photograph of UV (308 nm) laser ablation of polyimide. Ambient pressure is 1 atm air [5.53], R is the distance above the polymer surface

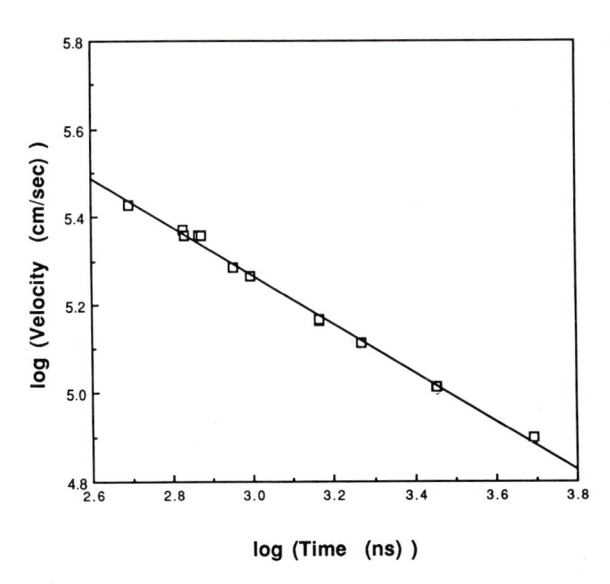

Fig. 5.16. Plot of the velocity of the blast wave vs time in the ablation of polyimide by UV photons

if reactive species are ejected from the polymer surface and undergo secondary reactions which are exothermic. In this connection, the species that are luminescent are of interest. The spectrum of the luminescent species has been analyzed [5.53] and found to consist of reactive diatomics such as CN, C_2^*, and CH. These would be expected to undergo reactions to form large molecules with the evolution of the heat of the process. The emitting species seem quite sensitive to the pressure of the ambient atmosphere as well as to the temporal behavior of the plume. The original work should be consulted for the details.

The black debris that is formed during UV-laser ablation of polyimide has been discussed before in Sect. 5.3.1. It is remarkable that it is scarcely visible in photographs such as those in Fig. 5.14. In two high-speed exposures [5.68] of the ablation of polyimide by a pulse (≈ 200 ns FWHM) of infrared-laser radiation at 9.17 μm from a CO_2 laser, the presence of a large amount of dense material is easily seen (Fig. 5.17). Since the presence of C_2^* and presumably C and C_3 as well) at distances of nearly a millimeter has been determined as described above [5.53], the polymerization of carbon to give large carbon clusters and solid carbon may occur in the ablation of polyimide in air on cold surfaces such as the polymer surface itself as the hemispherical envelope of the blast wave spreads laterally. Some of these particles would be sucked back into the etched area by the partial vacuum that is created behind the driving gas. When the ablation is conducted in a vacuum chamber, the black debris has been observed to coat the window through which the laser beam enters the chamber provided it is placed normal to the polymer surface [5.53].

b) Polymethyl Methacrylate

PMMA absorbs UV photons either poorly (248 nm) or moderately (193 nm). In contrast to polyimide, it is a linear, addition polymer which has one ester group

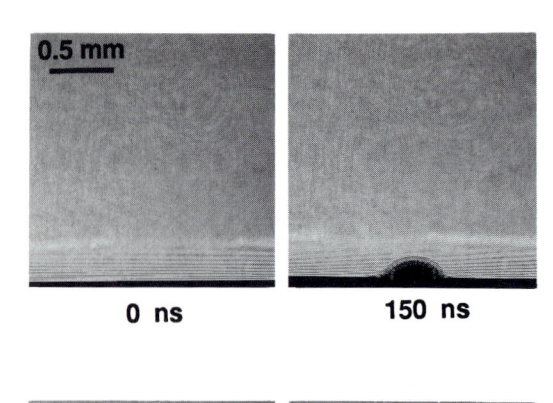

Fig. 5.17. Photographs of the blast wave rising above a polyimide surface during ablation by pulsed, CO_2-laser radiation at 9.17 μm [5.68]

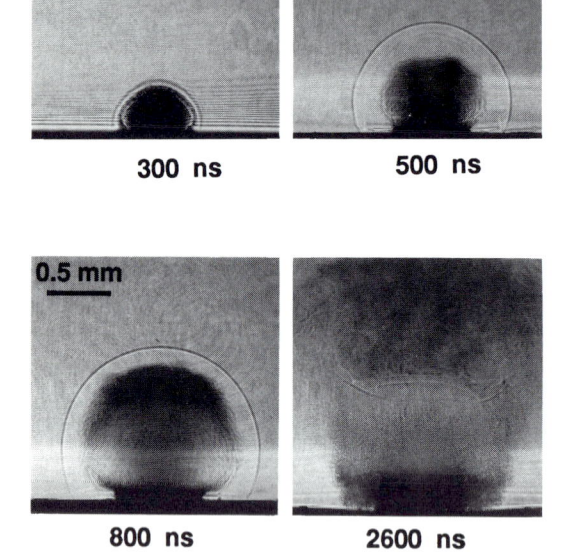

in the side chain of each monomer unit. These ester groups are the chromophores in the ultraviolet. The analytical work on its ablation products has been summarized [5.11]. At a laser wavelength of 193 nm, as much as 18% of the polymer that is ablated is accounted for as the monomer but the principal product is a low-molecular weight ($M_n < 2500$) solid material compared to an initial molecular weight of $\approx 10^6$. At the wavelength of 248 nm, there is only 1% of the monomer while the bulk of the material is ejected as a solid polymer of $M_n \approx 2500$.

The temporal changes in a PMMA film under the impact of a UV-laser pulse attracted the attention of researchers even as early as in 1986. As with polyimide [5.39], the stress wave that passed through the film as a result of the UV-laser pulse was monitored by a piezoelectric transducer. It was determined that at fluences above the threshold for ablation, the transducer signal was detectable with a delay of only a few nanoseconds after the start of the laser pulse. It suggested that the break-up of the polymer started even during the laser pulse. The possibility that the latter part of the laser pulse may be attenuated or hindered from normal absorption by the transformation of the polymer (in the optical path) by the leading part of the pulse was examined by means of optical transmission measurements [5.60]. It was found that the transmitted beam (for thin samples) was considerably attenuated during the laser pulse itself. A qualitative idea of the change in the reflectivity of the surface was obtained photographically using a fast (< 1 ns) visible-laser pulse as shown in Fig. 5.18 [5.49]. These changes have been measured quantitatively by the change in the width (FWHM) and the amplitude of the reflected pulse [5.56].

By timing the arrival of the products at a pyroelectric detector that was placed at a distance of several centimeters from the polymer film during ablation by a single 248 nm pulse (the exposure being performed in a vacuum [5.67]), it was determined that the gaseous products were expelled at supersonic velocities but the high-molecular weight ($M_n \approx 1000$) material travelled at subsonic velocities. The temporal profile of the ablation contrasted strongly with the behavior of polyimide in that the ejection of the products lasted over 800 μs. There was only one single maximum in the amplitude of the signal that was recorded by the detector.

As in the case of polyimide, it has proved difficult to give a complete picture of the process of ablation from results that have been obtained by various analytical methods. The problem has become even semantic because of the

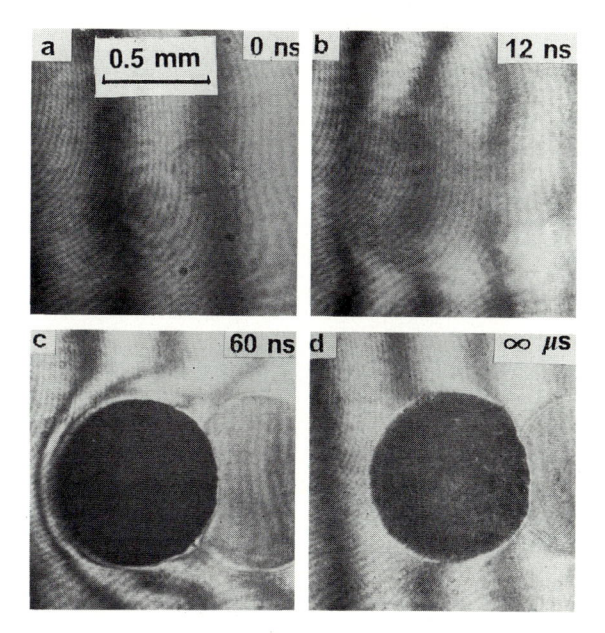

Fig. 5.18. Pictures of a PMMA surface during one single UV (248 nm; 20 ns FWHM) ablation pulse [5.49]

126   R. Srinivasan

introduction of new phrases by different groups to describe the observations. As an illustration, the series of photographs of the ablation of PMMA by a 248 nm laser pulse (Fig. 5.19) [5.68] can be offered. The laser fluence was selected to be sufficient to cause ablation even during the very first pulse and a fresh surface was used for each photograph. The emergence of the hemispherical blast wave is, as in the case of polyimide (Fig. 5.14) indicative of the lateral migration of the

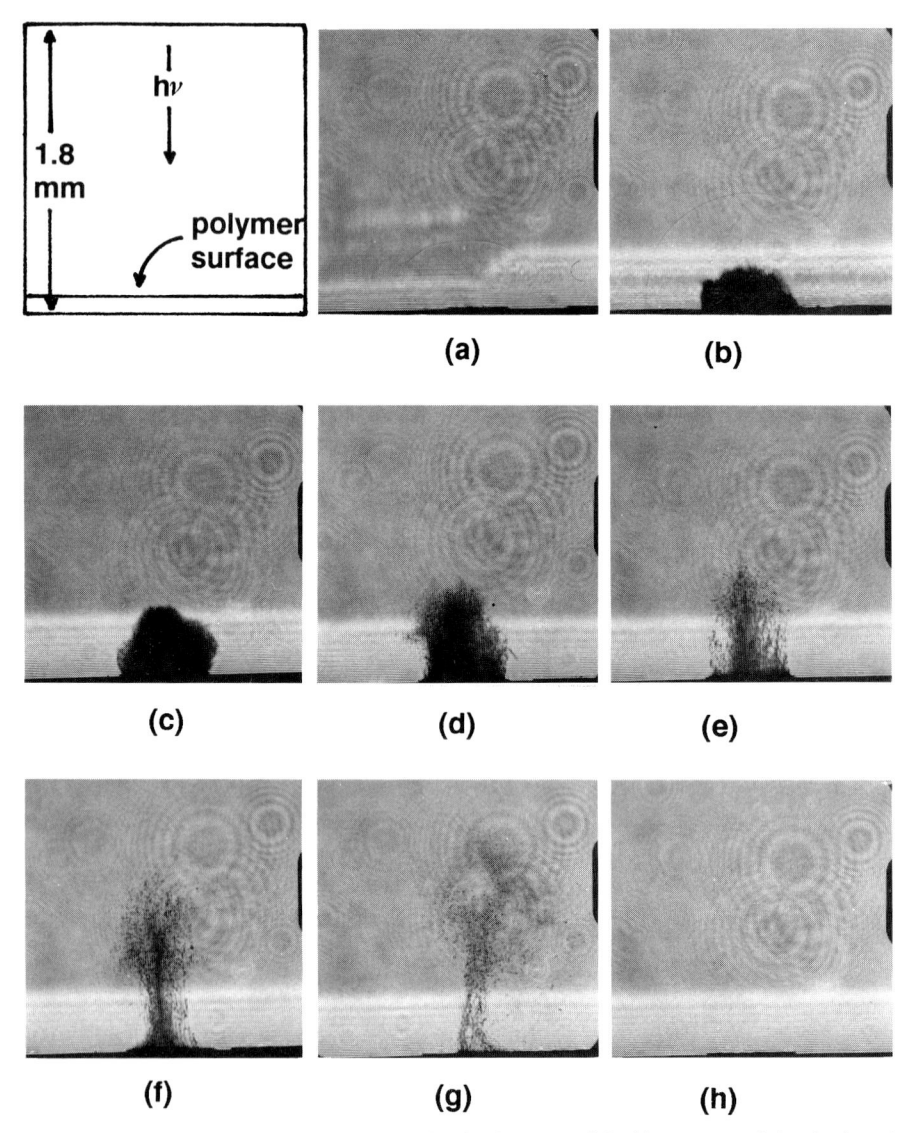

Fig. 5.19. Sequence of photographs showing the development of the blast wave and the ejection of dense material during the ablation of a PMMA surface by a single laser pulse (248 nm). (**a**) 750 ns; (**b**) 1.0 µs; (**c**) 3.0 µs; (**d**) 6.1 µs; (**e**) 9.7 µs; (**f**) 15.0 µs; (**g**) 20.7 µs; (**h**) ∞ s [5.68]

low-molecular weight product molecules through momentum exchange. The dense material that is first seen at ≈ 700 ns is undoubtedly the solid polymeric material that is the major product. This stream maintains its normal direction to the polymer surface until it is slowed down at a distance of ≈ 2 mm from the surface. Some puzzling conclusions have been drawn from the features of these photographs [5.69]. The dark column which is termed "ejecta" is said to suffer 'an abrupt loss of density from the target surface to roughly half the distance to the leading edge' and, to be 'imaged preferentially towards the axis as if an imaging threshold were playing a role'. The enlarged versions of Fig. 5.19 d–g are displayed in Fig. 5.20. They show that the dark stream of dense material undergoes a steady narrowing with time as if the orifice through which it is ejected is decreasing in diameter. This is seen to be correct on examining the surface by scanning electron microscopy at the end of the irradiation. This photograph has been published [5.14]. As the stream narrows, the dense material also melts and the enlarged photos (Fig. 5.20) show droplets in the stream until in Fig. 5.20g the stream turns into a jet of liquid for the most part. It

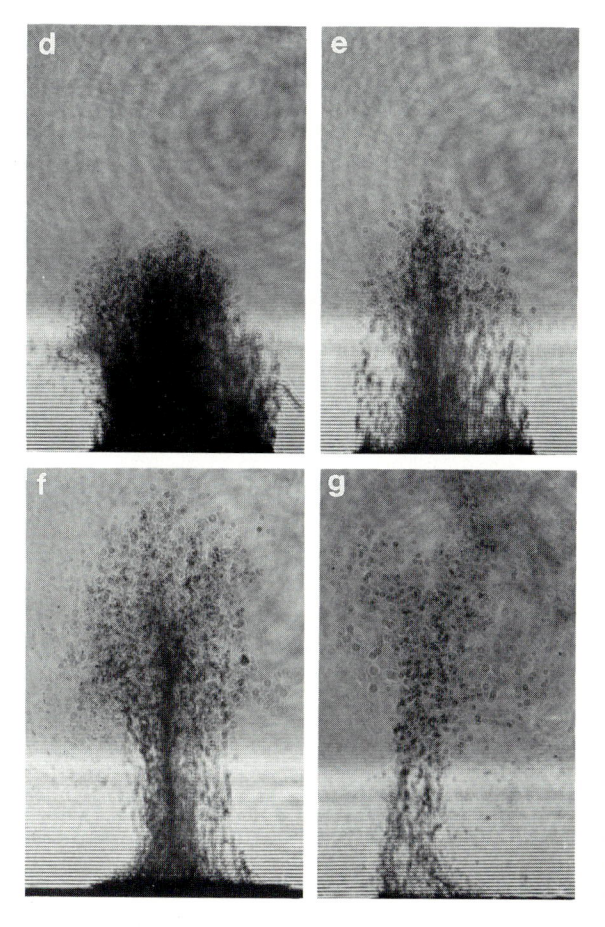

Fig. 5.20. Enlarged view of frames **d–g** from Fig. 5.19. In (**e**) the narrowing of the stream of dense material with time and in (**g**) the increasing amount of droplets can be seen [5.68]

suggests that the vibrational temperature of the initially solid material increases probably because the gas stream that propels it also increases in temperature. It should be remembered that the penetration depth of 248 nm photons in PMMA is > 20 µm while the etch depth per pulse is < 4 µm for the single pulse that was used. It means that the exposed region of the polymer film that underlies the ablated volume can have gaseous molecules such as CO and CO_2 trapped in it. This trapped gas will be released over a period of many microseconds as the surface first melts and then resolidifies. The bubbling of gases through the surface layer is evident in the SEM picture of the ablated surface [5.14]. An analysis [5.69, 70] of the gas dynamics of the laser-ablation process which does not take into account the chemistry of the reactions that the polymer undergoes on exposure to the laser photons is bound to be of limited value.

In this connection, it is useful to examine the photographs (Fig. 5.21) of the ablation of the same PMMA surface by pulsed (FWHM ~ 180 ns) infrared radiation at 9.17 µm from a CO_2 laser [5.68]. The interaction of CO_2-laser radiation with PMMA is known [5.71] to give the monomer, methyl methacrylate (b.p. 100.5 °C), as the only product. It is therefore not surprising that the photographs show a highly refractive vapor filling the hemispherical volume inside the blast wave. The contact front and its separation from the blast-wave front are clearly delineated. A comparison of these photographs to those from the ablation by pulsed, 248 nm laser radiation shows the sharp contrast between the chemical processes that drive the process of ablation in the two instances.

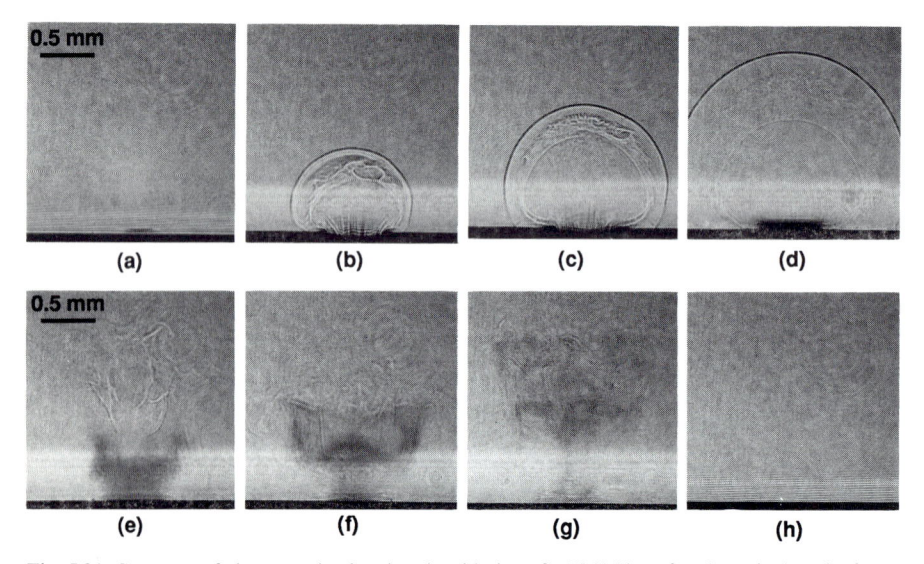

Fig. 5.21. Sequence of photographs showing the ablation of a PMMA surface by a single pulse from a CO_2 laser (9.17 µm) (**a**) 60 ns; (**b**) 350 ns; (**c**) 700 ns; (**d**) 1100 ns; (**e**) 3 µs; (**f**) 5 µs (**g**) 10 µs; (**h**) ∞ s [5.68]

5.4 Theories of Ultraviolet-Laser Ablation

The number of theories that have been proposed to explain the interaction of UV-laser pulses with polymer surfaces has been surprisingly large [5.6, 9, 13, 72–92]. Nearly all of them are based on an attempt to derive a relationship between the etch depth/pulse and fluence or log fluence and compare it to the experimentally observed relationship. For a given polymer, only the absorption coefficient is taken into account, although even that is considered unimportant in one theory [5.78]. The width of the laser pulse is usually not included. In more recent theories, the possibility that the plume that leaves the polymer surface is screening the photons that are coming in towards the latter part of the pulse is taken into account. The plume could consist of species which absorb more strongly or less strongly than the polymer itself [5.13, 79].

Apart from the objective given above for the proposal of many of these theories, a second objective is to gain an insight into the nature of the decomposition process by which the UV photons that are absorbed by the polymer chemically break up the material and cause it to ablate. In Fig. 5.5, the decomposition of the polymer is depicted as a step that can be separated from the ablation step. In practice, this may not be correct. The view that is prevalent is that in the depth l_f of the material that is etched by a single laser pulse, the polymeric bonds may be cleaved **and** the products removed by ablation even in the duration of the pulse. The spectra of the polymers that have been studied show that in the UV region, the absorption, at low fluences, corresponds to an electronic transition. A decomposition from this initially formed electronically excited state S_1 would constitute a photochemical decomposition. If the S_1 state undergoes internal conversion first, then a vibrationally "hot" ground-state chromophore would be produced which can lead to a photothermal decomposition process. There are complicating factors such as the crossing of the initially formed S_1 excited state (which is nearly always an upper singlet state in these polymers) to a triplet state T_1, which can also decompose or internally convert to the hot ground state. At power densities of 1 MW/cm^2 or more which are employed in UV-laser ablation, the S_1 or T_1 states – the latter being long-lived – can absorb a second photon and be promoted to an upper, S_2 or T_2 state which is bound to decompose rapidly to give a variety of small molecules. Such incoherent two-photon excitation and decomposition processes have been well-documented in small organic molecules [5.3]. Another complication is that as the duration of the laser pulse is finite, excited states that are internally converted during the first part of the pulse can provide the activation energy to decompose the excited states that are created by the later part of the pulse [5.41]. Such an activated photochemical decomposition would become particularly important as UV-laser pulses of longer (> 100 ns) duration are employed. At the other extreme, when UV-laser pulses of < 1 ps are used, it is quite possible that coherent, two-photon excitation of the polymer chromophores is involved in the ablation process. Thus, by the use of femtosecond UV-laser pulses, PMMA has been smoothly etched at 308 nm, a wavelength at which it is essentially transparent to a low-intensity UV beam [5.93]; a polymer such as Teflon™ (polytetrafluoroethylene) which cannot be etched with nanosecond pulses can be etched with femtosecond pulses [5.94].

It is disappointing that nearly ten years after the first reports on the interaction of UV-laser pulses with organic polymers, the theory and modelling of the phenomenon has been limited to one kind of data (the etch curves) which are obtained over a relatively narrow range of fluences and limited pulse widths. There are a few data of a quantitative nature that have been obtained with very short pulses as mentioned above but these are not usually covered by the theories. What happens to the laser pulses of longer (> 1 µs) duration which also seem capable of etching the polymer surface without any ablation [5.95] has never been predicted by any theory. Plots of log(etch depth/pulse) versus 1/fluence which were discussed in Sect. 5.2 suggest that the etching process that occurs with these 'long' pulses may have considerably larger activation energies than the processes which occur with ablation when nanosecond pulses are used [5.69]. Theories also do not address cases of doped polymers [5.96–99] whose etching behavior by UV-laser pulses can be enhanced relative to the undoped material even though the photons are certainly absorbed by the dopant. The obvious explanation of a photothermal process is ruled out by the fact that at two different UV wavelengths at which the doped material has an identical absorptivity, the etch curve is significantly different [5.97]. Theories also do not address the chemistry of the products that are formed by UV-laser ablation, especially the wavelength dependence that is seen in the composition of the products. In fact, the saddest criticism of UV-laser ablation studies to date is the meagerness of the analytical data on the products and the lack of a chemical mass balance in nearly all of the systems that have been studied.

5.5 Contemporary Trends in UV-Laser Ablation

The level of activity in this field has continued to increase from year to year over the past decade. It would be appropriate to conclude this review by pointing out several new directions in which the interaction of UV-laser radiation with polymers is being pursued. While the activity in any single area may be small at present, it shows the probable directions in which this field may develop over the next decade.

If the depth that is ablated from a polymer surface is minimized by controlling either the fluence of the UV laser or the number of pulses, it is possible to limit the etching to a mere change in the morphology of the surface. Such a modification of the surface has attracted considerable research interest [5.100–107]. The extent of the modification of the surface is a matter of definition. Changes which are no deeper than 50–100 Å are investigated by spectroscopic methods which were already mentioned in Sect. 2. Here we refer to changes that are far larger, of the order of 1–5 µm. Surface modification of fibers [5.101, 102, 106,107] and of polymeric surfaces in manufactured products [5.105] form a useful technology which can be expected to grow in importance.

If the entire thickness of a polymer film is < 1 µm, the modification of the surface as described before actually becomes an etching process. Such a controlled etching on a microscale has been applied to Langmuir-Blodgett films where the scaling down is one dimensional [5.108, 109] and to microspheres [5.110, 111] which are truly microscopic in three dimensions. The microspheres require handling in a light trap which is provided by a second laser during the ablation. Since the microspheres are suspended in a liquid and also enclose a liquid which is of different composition from the shell, this research is a truly innovative step which bridges many new techniques.

The use of UV-laser pulses of microseconds duration to bring about the etching of polymer surfaces without ablation [5.95] was mentioned in Sect. 5.4. The source of these pulses is a continuous wave (cw), UV laser whose beam is chopped mechanically to provide the pulses of 10–1000 µs duration. It is also possible to use the laser as a continuous source and translate the beam over the polymer surface in order to control the dwell time over an area that is equal to the spot diameter to be in the desired microsecond range. If the laser spot has a power density of 10–100 kW/cm^2, this technique provides a way to etch the surface with as little thermal damage as with nonosecond pulses from an excimer laser operating at > 1 MW/cm^2 of power density. This technique which has been called 'photokinetic etching' [5.112, 113] promises to extend the useful region in which UV-laser interactions with polymer surfaces can be studied and used.

References

5.1 J.G. Calvert, J.N. Pitts, Jr: *Photochemistry* (Wiley, New York 1966)
5.2 L. Zandee, R.B. Bernstein: J. Chem. Phys. **70**, 2574 (1979)
5.3 B.D. Koplitz, J. McVey: J. Phys. Chem. **89**, 4196 (1985)
5.4 R. Srinivasan, V. Mayne-Banton: Appl. Phys. Lett. **41**, 576 (1982)
5.5 R. Srinivasan, W.J. Leigh: J. Am. Chem. Soc. **104**, 6784 (1982)
5.6 T.F. Deutsch, M.W. Geis: J. Appl. Phys. **54**, 7201 (1983)
5.7 M.W. Geis, J.N. Randall, T.F. Deutsch, N.N. Efremow, J.P. Donnelly, J.D. Woodhouse: J. Vac. Sci. Technol. B **1**, 1178 (1983)
5.8 Y. Kawamura, K. Toyoda, S. Namba: Appl. Phys. Lett. **40**, 374 (1982)
5.9 J.E. Andrew, P.E. Dyer, D. Forster, P.H. Dey: Appl. Phys. Lett. **43**, 717 (1983)
5.10 S.T. Trokel, R. Srinivasan, B. Braren: Am. J. Ophthalmol. **96**, 710 (1983)
5.11 R. Srinivasan, B. Braren: Chem. Rev **89**, 1303 (1989)
5.12 P.E. Dyer: Laser ablation of polymers, in *Photochemical Processing of Electronic Materials* (Academic, London 1992) pp. 359
5.13 E. Sutcliffe, R. Srinivasan: J. Appl. Phys. **60**, 3315 (1986)
5.14 R. Srinivasan, B. Braren, K.G. Casey: J. Appl. Phys. **68**, 1842 (1990)

5.15 S. Kuper, M. Stuke: Appl. Phys. B **44**, 199 (1988)
5.16 S. Kuper, M. Stuke: Appl. Phys. A **49**, 211 (1989)
5.17 P.P. Van Sarloos, I.J. Constable: J. Appl. Phys. **68**, 377 (1990)
5.18 F. Bachmann: Chemtronics **4**, 149 (1989)
5.19 R. Srinivasan, B. Braren, R.W. Dreyfus: J. Appl. Phys. **61**, 372 (1987)
5.20 R. Srinivasan, B. Braren, D.E. Seeger, R.W. Dreyfus: Macromol. **19**, 916 (1986)
5.21 P. Dyer, J. Sidhu: J. Appl. Phys. **57**, 1420 (1985)
5.22 S. Lazare, R. Srinivasan: J. Phys. Chem. **90**, 2124 (1986)
5.23 M.C. Burrell, Y.S. Liu, H.S. Cole: J. Vac. Sci. Technol. A **4**, 2459 (1986)
5.24 A. Brezini: Phys. Status Solidii **127**, K9-13 (1991)
5.25 H. Niino, A. Yabe, S. Nagano, T. Miki: Appl. Phys. Lett. **54**, 2159 (1989)
5.26 H. Niino, M. Nakano, S. Nagano, A. Yabe, T. Miki: Appl. Phys. Lett. **55**, 510 (1989)
5.27 F. Kokai, H. Saito, T. Fujioka: Macromol. **23**, 674 (1990)
5.28 S.V. Babu, G.C. D'Couto, F.D. Egitto: J. Appl. Phys. **69**, 15 (1992)
5.29 H. Schmidt, J. Ihlemann, B. Wolff-Rottke: Appl. Surf. Sci. (in press)
5.30 R. Srinivasan, B. Braren, R.W. Dreyfus, D.E. Seeger: J. Opt. Soc. Am. B **3**, 785 (1986)
5.31 R.C. Estler, N.S. Nogar: Appl. Phys. Lett. **49**, 1175 (1986)
5.32 D.L. Singleton, G. Paraskevopoulos, R.S. Irwin: J. Appl. Phys. **66**, 3324 (1989)
5.33 DuPont Co., Polymer Products Dept., Kapton™ Polyimide Film Data Sheet (1983)
5.34 W.R. Creasy, J.T. Brenna: Chem Phys. **126**, 453 (1988)
5.35 C.E. Otis: Appl. Phys. B **49**, 455 (1989)
5.36 E.E.B. Campbell, G. Ulmer, K. Bues, I.V. Hertel: Appl. Phys. A **48**, 543 (1989)
5.37 E.E.B. Campbell, G. Ulmer, B. Hasselberger, I.V. Hertel: Appl. Surf. Sci. **43**, 346 (1989)
5.38 S. Kuper, J. Brannon: Appl. Phys. Lett. **60**, 1633 (1992)
5.39 P.E. Dyer, R. Srinivasan: Appl. Phys. Lett. **48**, 445 (1986)
5.40 R.S. Taylor, D.L. Singleton, G. Paraskevopoulos: Appl. Phys. Lett. **50**, 1779 (1987)
5.41 R.K. Al-Dhahir, P.E. Dyer, J. Sidhu, C. Foulkes-Williams, G.A. Oldershaw: Appl. Phys. B **49**, 435 (1989)
5.42 A.V. RaviKumar, G. Padmaja, P. Radhakrishnan, V.P.N. Nampoori, C.P.G. Vallabhan: Pramana **37**, 345 (1991)
5.43 P.E. Dyer, G.A. Oldershaw, J. Sidhu: J. Phys. Chem. **95**, 10004 (1991)
5.44 A.D. Zweig, T.F. Deutsch: Appl. Phys. B **54**, 76 (1992)
5.45 W.P. Leung, A.C. Tam: Appl. Phys. Lett. **60**, 23 (1992)
5.46 J.A. Sell, D.M. Heffelfinger, P.L.G. Ventzek, R.M. Gilgenbach: J.Appl. Phys. **69**, 1330 (1991)
5.47 E. Hunger, H. Pietsch, S. Petzoldt, E. Matthias: Appl. Surf. Sci. **54**, 227 (1992)
5.48 P. Simon: Appl. Phys. B **48**, 253 (1989)
5.49 R. Srinivasan, B. Braren, K.G. Casey, M. Yeh: Appl. Phys. Lett. **55**, 2790 (1989)
5.50 R. Srinivasan, K.G. Casey, B. Braren: Chemtronics **4**, 153 (1989)
5.51 R. Srinivasan, K.G. Casey, B. Braren, M. Yeh: J. Appl. Phys. **67**, 1604 (1990)
5.52 J. Grun, J. Stamper, C. Manka, J. Resnick, R. Burris, B.H. Ripin: Appl. Phys. Lett. **59**, 246 (1991)
5.53 P.E. Dyer, J. Sidhu: J. Appl. Phys. **64**, 4657 (1988)
5.54 P.L.G. Ventzek, R.M. Gilgenbach, J.A. Sell, D.M. Heffelfinger: J. Appl. Phys. **68**, 965 (1990)
5.55 D.L. Singleton, G. Paraskevopoulos, R.S. Taylor: Appl. Phys. B **50**, 227 (1990)
5.56 G. Paraskevopoulos, D.L. Singleton, R.S. Irwin, R.S. Taylor: J. Appl. Phys. **70**, 1938 (1991)
5.57 H.M. Pang, E.S. Yeung: Appl. Spectrosc. **44**, 1218 (1990)
5.58 M.N. Ediger, G.H. Petit: J. Appl. Phys. **70**, (1992)
5.59 M. Dienstbier, R. Benes, P. Rejfir, P. Sladky: Appl. Phys. B **51**, 137 (1990)
5.60 M. Golombok, M. Gower, S.J. Kirby, P.T. Rumsby: J. Appl. Phys. **61**, 1222 (1987)
5.61 J.K. Frisoli, Y.Hefetz, T.F. Deutsch: Appl. Phys. B **52**, 168 (1991)
5.62 G.H. Pettit, R. Sauerbrey: Appl. Phys. Lett. **58**, 793 (1991)
5.63 P.L.G. Ventzek, R.M. Gilgenbach, C.H. Ching, R.A. Lindley: J. Appl. Phys. **70**, (1992)
5.64 G. Koren, J.T.C. Yeh: Appl. Phys. Lett. **44**, 1112 (1984)
5.65 G. Koren, J.T.C. Yeh: J. Appl. Phys. **56**, 2120 (1984)

5.66   G.M. Davis, M.C. Gower, C. Fotakis, T. Efthimiopoulos, P. Argyrakis: Appl. Phys. A **36**, 27 (1985)
5.67   P.E. Dyer, R. Srinivasan: J. Appl. Phys. **66**, 2608 (1989)
5.68   R. Srinivasan: J. Appl. Phys. **73**, 2743 (1993)
5.69   B. Braren, K.G. Casey, R. Kelly: Nucl. Instrum. Methods B **58**, 463 (1991)
5.70   R. Kelly, A. Miotello, B. Braren, A. Gupta, K.G. Casey: Nucl. Instrum. Methods B **65**, 187 (1992)
5.71   M. Hertzberg, I.A. Zlochower: Combustion and Flame **84**, 15 (1991)
5.72   H.H.G. Jellinek, R. Srinivasan: J. Phys. Chem. **88**, 3048 (1984)
5.73   R.L. Melcher: In *Laser Processing and Diagnostics*. ed. by D. Bäuerle Springer Ser. Chem. Phys., Vol. 39 (Springer, Berlin, Heidelberg 1984) pp. 418
5.74   J.H. Brannon, J.R. Lankard, A.I. Baise, F. Burns, J. Kaufman: J. Appl. Phys. **58**, 2036 (1985)
5.75   T. Keyes, R.H. Clarke, J.M. Isner: J. Phys. Chem. **89**, 4194 (1985)
5.76   V. Srinivasan, M.A. Smrtic, S.V. Babu: J. Appl. Phys. **59**, 3861 (1986)
5.77   L.B. Kiss, P. Simon: Solid State Commun. **65**, 1253 (1988)
5.78   G.D. Mahan, H.S. Cole, Y.S. Liu, H.R. Philipp: Appl. Phys. Lett. **53**, 2377 (1988)
5.79   S. Lazare, V. Granier: Appl. Phys. Lett. **54**, 862 (1989)
5.80   R. Sauerbrey, G.H. Pettit: Appl. Phys. Lett. **55**, 421 (1989)
5.81   D.L. Singleton, G. Paraskevopoulos, R.S. Taylor: Chem. Phys. **144**, 415 (1990)
5.82   J. Singh, N. Itoh: Chem. Phys. **148**, 209 (1990)
5.83   K. Schildbach: In Proc. SPIE Int'l Congr. Opt. Sci. Eng., Paper 1279-07
5.84   M.S. Kitai, V.L. Popkov, V.A. Semchishen: SPIE **1352**, 210 (1990)
5.85   S. Kuper, S. Modaressi, M. Stuke: J. Phys. Chem. **94**, 7514 (1990)
5.86   G. Ya. Glauberman, S. Yu. Savanin, V.V. Shkunov, D.E. Shumov: Kvantovaya Elektron. (Moscow) **17**, 1054 (1990)
5.87   L.I. Kalantarov: Philos. Mag. Lett. **63**, 289 (1991)
5.88   G.A. Oldershaw: Chem. Phys. Lett. **186**, 23 (1991)
5.89   S. Etienne, J. Perez, R. Vassoile, P. Bourgin: J. Phys. III **1**, 1587 (1991)
5.90   N.P. Furzikov: SPIE **1503**, 231 (1991)
5.91   G. Huadong, G.A. Voth: J. Appl. Phys. **71**, 1415 (1992)
5.92   S.R. Cain, F.C. Burns, C.E. Otis: J. Appl. Phys. **71**, 4107 (1992)
5.93   R. Srinivasan, E. Sutcliffe, B. Braren: Appl. Phys. Lett. **51**, 1285 (1987)
5.94   S. Kuper, M. Stuke: Appl. Phys. Lett. **54**, 4 (1989)
5.95   R. Srinivasan: J. Appl. Phys. **72**, 1651 (1992)
5.96   H. Masuhara, H. Hiraoka, K. Domen: Macromol. **20**, 450 (1987)
5.97   R. Srinivasan, B. Braren: Appl. Phys. A **45**, 289 (1988)
5.98   H. Hiraoka, T.J. Chuang, H. Masuhara: J. Vac. Sci. Technol. B **6**, 463 (1988)
5.99   M. Bolle, K. Luther, J. Troe, J. Ihlemann, H. Gerhardt: Appl. Surf. Sci. **46**, 279 (1990)
5.100 H. Niino, M. Shimoyama, A. Yabe: Appl. Phys. Lett. **57**, 2368 (1990)
5.101 W. Kesting, T. Bahners, E. Schollmeyer: Appl. Surf. Sci. **46**, 326 (1990)
5.102 D. Knittel, E. Schollmeyer: Angew. Makromol. Chem. **178**, 143 (1990)
5.103 P.E. Dyer, R.J. Farley: Appl. Phys. Lett. **57**, 765 (1990)
5.104 E. Arenholz, V. Svorcik, T. Kefer, J. Heitz, D. Bäuerle: Appl. Phys. A **53**, 330 (1991)
5.105 P.T. Rumsby, M.C. Gower: Brit. Pat. 2,233,334 (1991)
5.106 T. Bahners, W. Kesting, E. Schollmeyer: SPIE **1503**, 206 (1991)
5.107 W. Kesting, E. Schollmeyer: Angew. Makromol. Chem. **193**, 169 (1991)
5.108 J.D. Magan, P. Lemoine, W. Blau, M. Hogan, D. Lupo, W. Prass, U. Scheunemann: Thin Solid Films **191**, 349 (1990)
5.109 S. Herminghaus, P. Leiderer: Appl. Phys. Lett. **58**, 352 (1991)
5.110 H. Misawa, M. Koshioka, K. Sasaki, N. Kitamura, H. Masuhara: Jpn. J. Appl. Phys. **30**, 907 (1991)
5.111 H. Misawa, N. Kitamura, H. Masuhara: J. Am. Chem. Soc. **113**, 2788 (1991)
5.112 R. Srinivasan: Appl. Phys. Lett. **58**, 2895 (1991)
5.113 R. Srinivasan: J. Appl. Phys. **70**, 7588 (1991)

6. Laser Ablation and Laser Desorption Techniques with Fourier-Transform Mass Spectrometry (FTMS)

R.L. Hettich and *C. Jin*

With 7 Figures

The advent of pulsed lasers has enabled a variety of experiments to be performed that had previously been impossible. In particular, the combination of lasers with mass spectrometers such as Time-Of-Flight (TOF) mass analyzers provides an experimental method for generating unusual species such as bare-metal clusters as well as desorbing large non-volatile biomolecules. One particular mass analyzer, a Fourier Transform ion cyclotron resonance Mass Spectrometer (FTMS), can easily be interfaced with lasers and has been demonstrated to be quite useful for these studies [6.1]. The extensive ion trapping and ion manipulation capabilities of FTMS can be used to examine the structures and reactivities of ions generated by the laser ablation and laser desorption processes. The focus of this chapter is not to extensively review the literature of laser applications with FTMS, but rather provide a broad overview of some of the types of experiments that can be performed with this technique. In particular, we will illustrate the basic principles of FTMS ion trapping and detection, and then focus on recent developments and applications of laser ablation FTMS for the study of cluster species and laser desorption FTMS for the examination of large biomolecules.

Because FTMS is a pulsed technique, this mass spectrometer can be directly interfaced with pulsed lasers. Ion detection for FTMS is accomplished without scanning, making it possible to acquire a complete mass spectrum from a single laser pulse, although multiple laser shots are usually co-added prior to Fourier transformation to improve the signal to noise ratio. While the theoretical mass range of FTMS is unlimited, the actual instrumental mass range of FTMS is approximately 50 000 amu for a 3 T magnet [6.2], making the combination of laser desorption FTMS quite attractive for the detection of large ions. Until recently, the most difficult aspect in evaluating the useable mass range of FTMS was the ability to *create* large ions. Laser desorption methods can now be used to generate species with masses exceeding 100 000 amu [6.3], so that ion trapping and/or detection factors can now be examined. Although the mass range of conventional time-of-flight mass spectrometers is greater than 500 000 amu, the FTMS technique offers a number of advantages over TOF mass spectrometry. First, the accurate mass and ultrahigh-resolution mass measurement capabilities of FTMS provide the ability to definitively identify the empirical formula of a given ion and resolve isobaric species. Second, the ion trapping and ion manipulation capabilities of the FTMS instrument provide a means of probing ionic structures in great detail, and often provide information which can be used to resolve structural isomers. In particular, the ability to probe ion-molecule reactions and various fragmentation processes such as collision-activated dissociation and photodissociation can be used to extensively interrogate ionic structures, and will be demonstrated below.

6.1 Principles of FTMS Operation

6.1.1 Ion Formation

Fourier transform mass spectrometry is an ion trapping technique that differs substantially from other types of mass spectrometers. The fundamental principles of FTMS have been reviewed [6.4]; the reader is referred to this article for more detailed information on the theory and practical aspects of FTMS. The basic hardware for FTMS consists of an analyzer cell (Fig. 6.1) contained within a UltraHigh Vacuum (UHV) chamber (pressure $< 10^{-8}$ Torr), which is centered in a strong, homogeneous magnetic field. Ionization can be accomplished in FTMS by a variety of methods, although we will focus discussion on laser processes. In a typical experiment, a sample is placed on a stainless-steel probe tip, which is then inserted into the vacuum chamber and placed near the FTMS ion cell. The sample is desorbed and ionized by a focussed pulsed laser beam (approximately 300 μm diameter spot size at power densities of 10^6–10^{10} W/cm^2). This event generates a variety of species from the sample, including ions (both positive and negative), neutral molecules and electrons. Several different types of lasers have been used with FTMS, the most common of which are the pulsed CO_2 (10.6 μm) and the pulsed Nd:YAG (1.06 μm) laser. Frequency mixing and doubling of the Nd:YAG laser can be used to produce radiation at 532 nm, 355 nm and 266 nm. These lasers provide a variety of

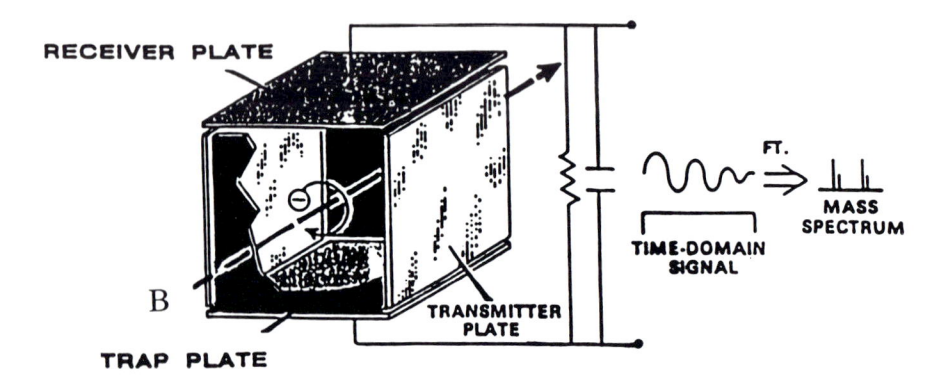

Fig. 6.1. The FTMS analyzer cell, identifying the two receiver plates, the two transmitter plates, the two trapping plates and the direction of the magnetic field (B). The time-domain image current is monitored between the two receiver plates and Fourier transformed to yield a mass spectrum

wavelengths, pulse widths and power densities. The laser is usually triggered to fire during the beam event of the mass spectrometer experiment sequence. The laser wavelength is usually less important than the laser power density for these desorption experiments; however, the laser wavelength can be chosen to match the spectral absorption region of a compound of interest in order to provide enhanced detection of a certain species. Laser ablation, which will be defined in this context as power densities $> 10^9$ W/cm^2, is useful for atomization and subsequent gas-phase synthesis of unusual species such as bare clusters and will be discussed in detail in the following sections. Laser desorption, which will be used to describe the 10^6–10^9 W/cm^2 power-density region, is more useful for the gentle volatilization of a sample and does not completely atomize the analyte of interest. This technique is suitable for the examination of non-volatile compounds such as large biomolecules which are present in a sample.

6.1.2 Ion Trapping

Once formed, ions can be confined inside the FTMS ion cell by a combination of electrostatic and magnetic forces. Ions in the FTMS analyzer cell are constrained by the magnetic field to move in circular orbits with motion confined perpendicular to the magnetic field (x-y plane) due to the Lorentz force, but not restricted parallel to the magnetic field (z-axis). Ion trapping along the z-axis is accomplished by applying an electrostatic potential to the two plates on the ends of the cell, shown in Fig. 6.1. The trapped ions can be stored in the FTMS analyzer cell for long periods of time (up to hours), provided that a high vacuum is maintained to reduce the number of destabilizing collisions between ions and residual neutral molecules. The laser creates positive ions and negative ions simultaneously; the polarity of the voltage applied to the trapping plates determines whether positive or negative ions are retained in the cell. The ion motion in the cell is complex due to the presence of electrostatic and magnetic trapping fields and consists of three different modes of oscillations [6.5]; however, the primary mode of interest is the cyclotron motion, whose frequency v_c is directly proportional to the strength of the magnetic field B (typically between 3 and 7 T) and inversely proportional to the mass to charge ratio m/z of the ions, as shown in equation (6.1)

$$v_c = \frac{kzB}{m}, \tag{6.1}$$

where k is a proportionality constant.

As may be seen in this equation, ions of different m/z have unique cyclotron frequencies. For example, at a magnetic field strength of 3 T, an ion of m/z 40 will have a cyclotron frequency of 1.2 MHz, while an ion at m/z 5000 will have a cyclotron frequency of 9.4 kHz. Equation (6.1) also reveals that increasing the magnetic field linearly increases the cyclotron frequency of ions, making high-mass ions easier to detect over environmental noise that exists in the low kHz region. Additional benefits of increasing the magnetic field include improvement in mass-resolving power, increased ion-trapping efficiency (and thus better signal to noise ratio) and an extension of the upper mass limit [6.4]. These are some of the major reasons why many FTMS instruments are being upgraded with 7 T magnets rather than the more conventional 3 T magnets.

6.1.3 Ion Detection

Unlike mass spectrometers that employ electron multipliers for ion detection, FTMS ion detection is accomplished by monitoring the image current induced by the orbiting ion packet as it cycles between the two receiver plates. After formation by an ionization event, all trapped ions of a given mass-to-charge ratio have the same cyclotron frequency, but have random positions about the center axis of the FTMS cell. The net motion of the ions under these conditions does *not* generate a signal on the receiver plates of the FTMS analyzer cell, shown in Fig. 6.1, because of the random locations of ions and their distance from the receiver plates. In order to detect cyclotron motion, an excitation pulse must be applied to the FTMS cell to make the ions "bunch" together spatially into a coherently orbiting ion packet. This excitation pulse also increases the radius of the orbiting ion packet so that it closely approaches the receiver plates of the FTMS cell. As a result, the net coherent ion motion produces a time-dependent signal between the receiver plates, termed "image current", which is representative of all the ions present in the FTMS cell. Experimentally, this induction of coherent motion is accomplished by applying a broadband rf "chirp" pulse, a well-defined SWIFT (Stored Waveform Inverse Fourier Transform) pulse, or a voltage impulse spike [6.5]. The amplitude of the excitation pulse must be carefully controlled for ion detection to avoid exciting the ions to such a large radius that they collide with the cell plates and are annihilated.

This image current is converted to a voltage, amplified, digitized, and Fourier transformed to yield a frequency spectrum, which contains complete information about frequencies and abundances of all ions trapped in the cell. Note that the FTMS detection scheme does not require scanning the magnetic field, but that the induced image current is a simultaneous summation of the coherent motion of all ions present in the FTMS cell. A mass spectrum can then be determined by converting frequency into mass via equation (6.1). Because frequency can be measured precisely, the mass of an ion can also be determined very precisely, to one part in 10^9 or better. Experimentally, the trapping-plate voltages as well as the number of ions present in the cell can affect accurate mass measurements and must be carefully controlled. Unlike other mass-spectral techniques which require many calibrant ions to completely span the mass range of interest, mass calibration in FTMS can be accomplished with only a few calibrant ions, preferably at masses above and below the region of interest.

In a conventional 2-inch-cubic FTMS cell, as few as ten ions can be detected; however, when more than 10^6 ions are present, ion–ion repulsion begins to significantly degrade the trapping process and results in a substantial deterioration of the ion signal and resolution. This space-charge limit defines the dynamic range of the FTMS (approximately 10^5). By ejecting the most abundant ions prior to detection, this dynamic range can be increased for the detection of minor components in a mixture. FTMS ion detection is non-destructive and can be repeated many times to improve signal to noise ratio because the ion signal increases as the square-root of the number of detection events. In a typical FTMS experiment, after ion detection, a voltage spike is used to eject all ions from the cell in preparation for a new experiment. However, instead of ejecting the ions from the cell, they can be allowed to relax to the center of the cell through collisions with background gases. These ions can then be re-excited and measured several times [6.6]. This remeasurement process has the potential of reducing the amount of sample required for FTMS analysis, which is particularly important in biological applications. Even though FTMS detection of image current is somewhat less sensitive than ion detection with an electron multiplier, the remeasurement technique provides a method of enhancing the FTMS detection process, perhaps even to the point of detecting a single ion.

Since mass resolution is proportional to the time the image current can be monitored, it is advantageous to keep the background gas pressure in the FTMS cell as low as possible (below 10^{-8} Torr) in order to reduce the number of collisions that occur. Collisions between the orbiting coherent ion packet and the neutral molecules in the cell result in a loss of coherent motion and a reduction in the time that the image current can be detected above the noise. While low- to medium-resolution ($m/\Delta m < 20\,000$) FTMS experiments require a few milli-seconds to acquire a complete mass spectrum, high-resolution ($m/\Delta m > 20\,000$) measurements are achieved by adjusting the experimental conditions so that the image current can be acquired for times as long as several seconds. Because of the inverse relationship between frequency and m/z in equation (6.1), it is more difficult to resolve high-mass (low-frequency) ions than low-mass (high-frequency) ions. Whereas ultrahigh mass resolution can be achieved with a 3 T magnet for small ions ($m/\Delta m > 1\,000\,000$ for m/z 100), FTMS resolution decreases linearly with increasing mass, indicating that a resolution of approximately $m/\Delta m$ 10 000 would be expected for an ion with m/z 10 000 under the same experimental conditions.

6.1.4 Ion Structural Techniques

The most attractive features of the FTMS technique are its extensive ion trapping and manipulation capabilities, which provide a mechanism for probing ionic structures in great detail. By applying rf pulses to the FTMS cell which correspond to the cyclotron frequencies of the ions, it is possible to excite these ions to a larger radius of motion and therefore a higher translational energy. By carefully controlling the amplitude (or energy) of the rf pulse, it is possible to conduct different experiments. Low-medium-energy excitation is used to accelerate selected ions into a collision gas to induce fragmentation, as outlined later. Medium-energy excitation is used to excite the ions into coherent motion for ion detection, as outlined before. High-energy excitation is used to excite the ions to such large radii that they collide with the FTMS cell plates and are annihilated. By controlling which rf frequencies are applied to the FTMS cell, it is possible to selectively eject unwanted ions from the cell.

After this ion-isolation step, a number of processes can be conducted to probe the structure of these trapped ions using "reagents" such as neutral molecules, photons and electrons. For example, selected ions can be controllably accelerated into a neutral gas, such as argon, to produce fragment ions. This process, called Collisionally Activated Dissociation (CAD) or tandem-mass spectrometry, can provide detailed structural information [6.7]. Further, multiple steps of ion isolation and subsequent dissociation can be linked together to permit the investigation of several "generations" of fragment ions. This process is referred to as MS^n (where n is the number of stages of tandem-mass spectrometry), and is readily achievable on trapped-ion instruments such as FTMS or rf-quadrupole ion-trap mass spectrometers. For comparison, transmission-based instruments such as quadrupoles and magnetic sectors require additional hardware to obtain multiple stages of tandem-mass spectrometry.

As an alternative to collisional dissociation processes, photo-induced fragmentation can also be used with FTMS. In this process, after the initial isolation step, the trapped ions are irradiated with photons to induce fragmentation. This dissociation technique may provide different fragmentation information than CAD and is often more useful than CAD for the differentiation of isomers [6.8]. Ion-molecule reactions can be conducted by trapping the ions in the presence of various reagent gases and have also been used extensively with FTMS for probing the reactivities and structures of trapped ions. By monitoring the charge-exchange reactions between selected ions and neutral compounds, it is possible to bracket electron affinities and ionization potentials. Alternatively, proton transfer can be monitored between ions and neutrals, and provides a means of examining gas-phase acidities or basicities. The use of these structural techniques will be demonstrated with specific examples in the following sections.

6.2 Laser-Ablation FTMS for Clusters

6.2.1 Cluster Formation

Clusters are aggregates of atoms which bridge the region between molecules and bulk solids. The interest in the physics and chemistry of clusters has grown rapidly in recent decades [6.9], partially driven by the desire to use them as molecular models of bulk surfaces [6.10–13] and novel materials [6.14]. FTMS offers an unique way to study clusters in the gas phase. Compared with the more conventional time-of-fight mass analyzer, FTMS has extremely high mass resolution making it easy to distinguish clusters of different size and their reaction products. Because of the long ion-trapping time of the FTMS, clusters can be collisionally cooled or manipulated prior to examination of their reactivities.

One of the standard methods of gas-phase cluster generation is laser ablation (also called laser vaporization). An intense pulsed-laser beam is focussed on the sample and heats it locally to ultrahigh temperatures to vaporize the sample material into a hot plasma [6.15]. For some elements that easily aggregate, such as sulfur, ejected from a solid sample as molecular species, small clusters can be easily produced by direct laser ablation of the solid sample into high vacuum. Small silicon clusters, [6.16–18] and carbon clusters, including large fullerenes [6.19–22], have been produced by direct laser ablation near a FTMS cell under high-vacuum conditions.

For a typical metal, the laser plasma is composed mostly of atomic species. Due to their weak bonding forces, metal clusters are not as easily formed by direct laser ablation as are silicon and carbon clusters. A few exceptions to this rule have been observed. For example, aluminum cluster anions were generated by direct laser ablation of an aluminum disk in a high-vacuum FTMS chamber with no collision gas [6.23]. In this case, an aluminum disk was placed on a solid probe which was inserted into high vacuum and placed just outside the source cell of the FTMS instrument. The 1064-nm radiation of a Nd:YAG laser was focussed onto the aluminum disk and the resulting clusters anions were trapped and detected in the source cell. As shown in Fig. 6.2, cluster anions ranging form Al_3^- to Al_{50}^- were produced.

In general, direct laser ablation of a solid sample in a high vacuum is a simple and easy way to produce some types of clusters. It is compatible with the high-vacuum requirement of the FTMS and does not require any additional complicated differential pumping and ion optics. However, formation of larger clusters usually involves sequential fusion of two smaller species, and requires third-body collisions to remove the extra energy produced by the fusion process in order to stabilize the resulting clusters. To generate larger clusters by direct laser ablation in high vacuum, a dense laser plasma has to be formed to provide stabilizing collisions. A dense plasma can be produced by drilling a microcavity in the sample [6.19] or utilizing the magnetic confinement of the laser plume in the strong magnetic field near the FTMS cell [6.21].

The variety of clusters that can be produced by direct laser ablation in high vacuum is quite limited because of the lack of stabilizing third-body collisions. A better means of producing clusters is to perform the laser ablation of a solid sample in the presence of a buffer gas. The buffer gas confines the laser-plasma plume and provides the third-body collisions required to stabilize the larger clusters. A broad range of clusters of practically any material have been produced using this method [6.24–27]. However, the presence of the high-pressure buffer gas is incompatible with the high-vacuum requirements of the FTMS and therefore cannot be performed close to the FTMS cell. In order to address these concerns, researchers have developed an external cluster source which can be attached to the FTMS [6.28]. In this case, clusters are generated in

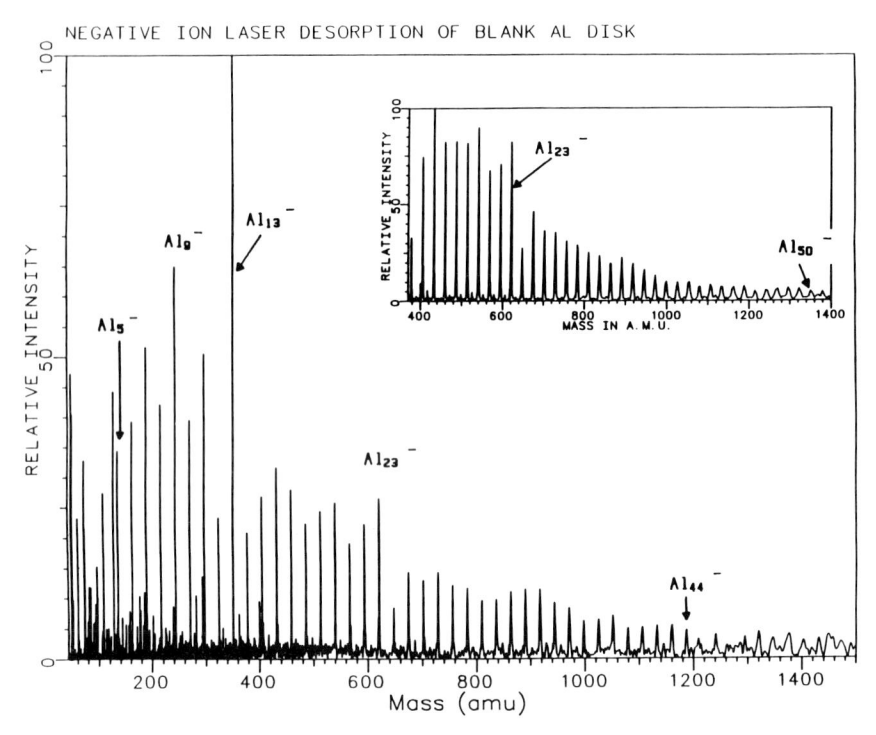

Fig. 6.2. The laser-ablation negative-ion FTMS spectrum of aluminium cluster anions generated from an aluminum target. Clusters ranging from Al_3^- to Al_{50}^- are observed, with noticeable differences in abundance. The insert shows an expansion of the high mass region. Adapted from [6.23]

an external chamber outside the high-vacuum region by laser ablation (vaporization) into a helium buffer gas. Ions are cooled by supersonic expansion into the high-vacuum region, injected into the magnetic field by a series of elaborate ion optics, and finally trapped in the FTMS cell. This technique was recently improved by constructing a compact cluster source and direct external injection [6.29]. With this external laser-ablation source, clusters of nearly any element can be generated and studied with the FTMS technique.

Small homonuclear and heteronuclear metal clusters can also be generated by ion–molecule reactions. For example, $Fe(CO)_5$ has been reacted with a variety of metal ions M^+ produced by laser ablation in the FTMS cell to produce clusters of MFe^+ [6.30]. The chemical reactivities [6.31, 32] and spectroscopy [6.33] of these species have been studied in detail and provide information about the bonding and properties of these small diatomic metal clusters. It is interesting to note that the chemistry of these mixed clusters is quite different from the chemistry of the individual metals themselves. For example, the reactivity of VFe^+ is substantially different from that of Fe^+ or V^+ [6.34], and implies that even these small-cluster ions behave quite differently than the bare-metal ions. Examination of the photodissociation spectra for these clusters and other organometallic species provided both thermodynamic and spectroscopic information, yielding bond energies, fragmentation products and absorption information [6.35].

Fullerenes, which are large cage-shaped carbon clusters consisting of an even number of carbon atoms, can be generated by laser ablation of a variety of carbon-rich materials, including graphite, coal, diamond and organic polymers. In order to detect fullerenes in a variety of samples, it is necessary to differentiate between fullerenes that were *generated* by the laser ablation event and fullerenes that were *desorbed* from the sample. One way of addressing this issue is to carefully control the laser-power density. Fullerenes are usually created by higher power density laser ablation ($> 10^9$ W/cm^2) of a carbon-containing material, whereas lower laser-power densities ($< 10^7$ W/cm^2) are usually insufficient to form fullerenes from these samples. This can be verified by examining graphitic samples which are known to be free of fullerenes under a variety of laser energies.

This experimental approach can be demonstrated by summarizing the experiments that were conducted to search for fullerenes in natural geological samples [6.36]. Electron microscopy of a carbon-rich rock from Russia, labelled "Shungite", revealed images that were very similar to images from synthetic fullerene samples. In order to definitively verify the presence of fullerenes in the sample, mass spectrometry was needed. Laser desorption (10^6 W/cm^2) FTMS revealed the presence of C_{60} and C_{70} in these Shungite samples. The Shungite samples and several blanks (graphitic material known to be free of fullerenes) were examined by this technique in a "blind" test in which the identities of the samples and the blanks were not disclosed until after the mass spectrometry had been performed. Fullerenes were detected only in the Shungite sample and not from any of the blanks under these experimental conditions. In order to further confirm that fullerenes were, in fact, present in the Shungite sample and not generated by the laser desorption event, thermal desorption FTMS experiments were conducted. For these experiments, samples of the Shungite rock and blanks were thermally desorbed from a stainless-steel probe tip at temperatures up to 350 °C. This temperature should volatilize the C_{60}, which could then be ionized by low-energy electron capture. Fullerenes cannot be generated from graphitic material under these experimental conditions and would not be observed if they were not present in the sample. These thermal desorption experiments verified the presence of fullerenes in the Shungite sample and not in the blanks, supporting the laser desorption experiments. As should be obvious from these studies, laser desorption is quite useful for the detection of fullerenes in samples; however, the use of high laser irradiance can distort the analysis and may create clusters which were not present in the original sample.

6.2.2 Accurate Mass and High-Resolution Measurements

The ultrahigh mass resolution of the FTMS provides an ideal way to resolve and identify isobaric clusters. Recently boron-doped Buckminsterfullerene clusters

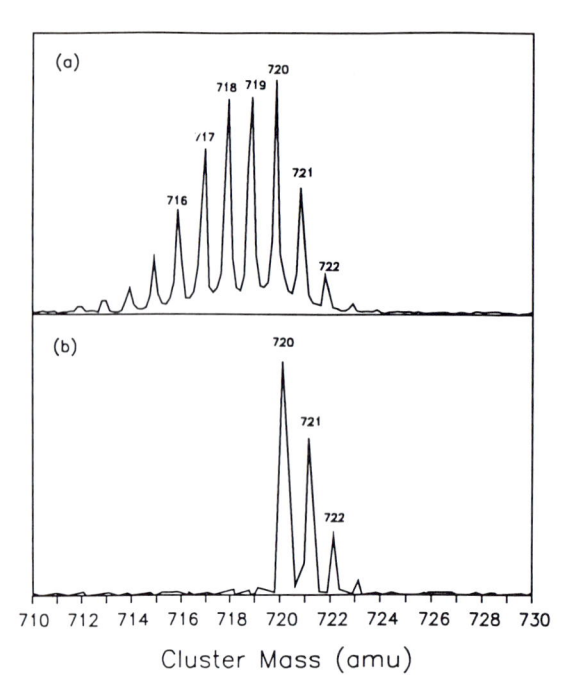

Fig. 6.3. FTMS positive-ion mass spectrum of clusters produced by laser ablation of (**a**) a boron nitride/graphite composite disk, and (**b**) a pure graphite disk. The peaks in (**b**) are due to carbon-only fullerenes whereas the peaks in (**a**) are due to 60 atom pure-carbon fullerenes and fullerenes in which up to four carbon atoms have been replaced with boron atoms

[6.37] were generated by laser ablating a boron-nitride/graphite composite disk in helium buffer gas and mass analyzed in a FTMS. The mass spectrum of the 60-atom cluster is shown in Fig. 6.3a. Mass peaks at 720 amu, 721 amu and 722 amu are due to pure C_{60} isotope distribution of $^{12}C_{60}$, $^{12}C_{59}{}^{13}C$ and $^{12}C_{58}{}^{13}C_2$, respectively. There are also cluster ions clearly present in significant abundance at 719, 718, 717, 716 and 715 amu. They correspond to Buckminsterfullerene cages in which up to four carbon atoms have been replaced by boron atoms (i.e., $C_{59}B$, $C_{58}B_2$, etc). The high mass resolution of the FTMS was necessary to identify these new molecules. For comparison, the mass spectrum from the laser ablation of a pure graphite disk is shown in Fig. 6.3b.

High resolution, accurate mass measurements were also used to identify multiply charged anions of C_{60} in a laser-desorption FTMS experiment [6.38, 39]. The ^{13}C isotope packets for the C_{60}^{2-} and C_{70}^{2-} ions were observed at half-mass units and in the correct abundances to identify these unusual ions, which were thought to be generated from fullerene samples by laser-induced surface ionization. Various ion-trapping and ion-ejection experiments were also used to identify these ions. Pseudopotential calculations indicated that C_{60}^{2-} is slightly bound (electron affinity of C_{60}^- was estimated to be 0.1–0.4 eV).

6.2.3 Ion–Molecule Reactions

Cluster ions can be trapped in the FTMS cell for a long time. This process allows introduction of various reaction gases into the cell to study the gas-phase ion–molecule reactions of the cluster ions. For example, aluminum-cluster anions produced by direct laser ablation in the FTMS apparatus (Fig. 6.2) were unreactive with most small molecules, except for oxygen [6.23]. Al_n^- ($n = 3-8$) would react with O_2 to produce both AlO_2^- and AlO^-. For the larger aluminum-cluster anions, the odd-numbered clusters were observed to be unreactive with oxygen, whereas the even-numbered clusters were observed to react rapidly with oxygen to generate the same ionic products observed for the small aluminum clusters. This odd–even alternation in reactivity was thought to be due to the stability of the neutral aluminum-cluster products generated by the reaction with oxygen.

Electron Affinities (EA) and Ionization Potentials (IP) can also be measured for clusters with the FTMS technique. For aluminum clusters, each anion was isolated and allowed to react with either nitrogen dioxide (EA = 2.30 ± 0.1 eV) or p-benzoquinone (EA = 1.89 ± 0.1 eV) [6.23]. The resulting bracketed electron affinities for these clusters were in good agreement with theoretical calculations and previous photodetachment-experiment results. Using similar charge-exchange methods, the ionization potentials of fullerenes were bracketed in the FTMS cell [6.40]. These experiments revealed that the first ionization potentials of the "magic number" clusters (i.e. C_{50}, C_{60} and C_{70}) were anomalously higher than the other neighboring clusters, indicating the stability of both neutral and ionic forms of these special clusters.

The boron cage doped fullerenes shown in Fig. 6.4a were found to be Lewis acids in which each boron atom would react with one ammonia molecule in the FTMS cell, as shown in Fig. 6.4b [6.37]. These results verified that the boron atoms were present as a part of the fullerene cage and not located inside the fullerene ball. The ion trapping and high mass resolution of the FTMS made these measurements possible and provided insight into the structure of the doped fullerenes and the location of the boron atoms.

The structures and internal temperatures of the clusters generated by laser ablation (vaporization) as well as other methods are generally unknown. This makes the experimental results difficult to control and causes variation from one instrument to another. The long ion-trapping time of the FTMS provides a means to cool cluster ions to the desired temperature and structure. Collision gases may be introduced into the cell while cooling or heating the trapping cell. After an extended period of colliding with the thermalized gas, the cluster ions are brought to equilibrium with the cell temperature. Cluster ions can also be manipulated to their energetically most favorable geometric structure. For instance, Ta_{10}^+ cluster ions were generated by laser ablation and trapped in the FTMS cell for examination of their reactivity with H_2 [6.41]. For Ta_{10}^+ cluster ions directly injected from an external laser ablation (vaporization) source, two reaction rate constants were obtained indicating the presence of at least two geometric isomers of Ta_{10}^+ in the cell. Borrowing the concept of "annealing" used in experimental surface sciences, where a relatively rough bulk surface is heated to high temperature and then is cooled down slowly to obtain a perfect single crystalline surface, Ta_{10}^+ cations were annealed by first heating them with laser

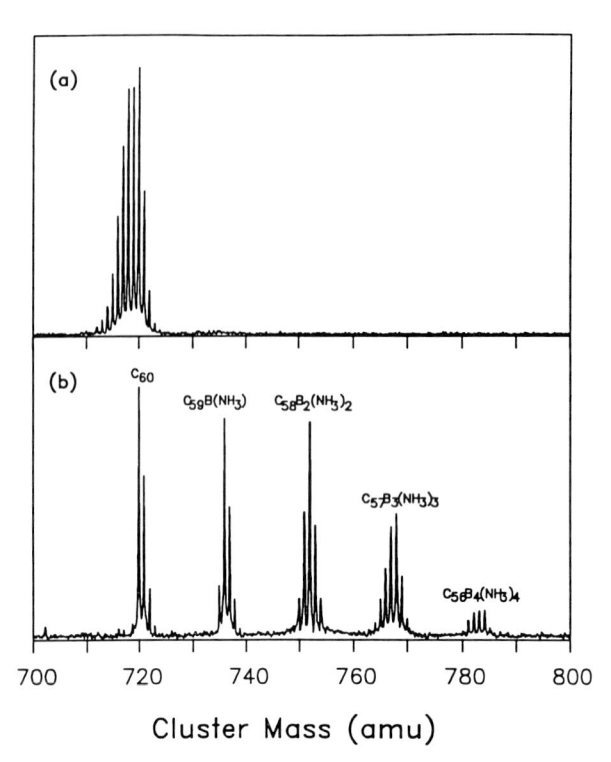

Fig. 6.4. Ion-molecule reactions of boron-doped fullerenes with ammonia. (**a**) The positive-ion FTMS spectrum of these cluster ions before reaction; (**b**) The 2 second reactions of these clusters with ammonia. Note that the number of ammonia molecules added to the clusters coincides exactly with the number of boron atoms in the doped fullerenes. Adapted from [6.37]

irradiation and then cooled to room temperature by collisions with thermal argon gas. Only one rate constant was obtained after annealing, suggesting that all Ta_{10}^+ cations were transformed to the most stable geometric structure.

6.2.4 Collision-Activated Dissociation

The fragmentation pathways of cluster ions trapped in the cell can be examined by controllably exciting and colliding these ions into an inert gas such as argon. This process generates fragment ions which aid not only in investigating the structures of the ions but also the energetics and pathways of dissociation. For instance, CAD studies revealed that the fragmentation of aluminum-cluster anions is a competition between electron detachment and aluminum-atom elimination [6.23]. This indicates that the bond energies of the aluminum atoms in a cluster anion are roughly equal to the electron affinities of these species.

The stability of CO chemisorption on Cu^+ clusters has been theoretically calculated [6.42]. Theory predicts that this metal/adsorbate system should behave according to the cluster-jellium model and Cu_7^+CO should be most stable in that size range. Using CAD methods in an FTMS, this theoretical prediction was tested by colliding the Cu_n^+CO and Au_n^+CO cations with argon [6.42]. All of these ions fragmented by elimination of CO, and Cu_7^+CO and Au_7^+CO were found to be the most stable species. This study is a good example of the use of clusters as molecular models of real bulk surfaces and shows the good agreement between cluster theory and experiment.

6.3 Laser-Desorption FTMS for Biomolecules

The examination of large, polar, non-volatile compounds such as proteins, peptides and oligonucleotides requires ionization techniques which can provide "soft" desorption and ionization conditions in order to minimize fragmentation. Lasers provide a rapid thermal spike which efficiently desorbs large molecules from a surface, and thereby eliminates the need to derivatize the biomolecules for increased volatility prior to mass analysis. However, the laser wavelength and power density are quite critical and often prohibit successful desorption of large molecules without inducing fragmentation. This results from absorption of either the laser radiation or the resulting thermal energy by the biomolecule, and results in fragmentation. Direct laser desorption, as outlined in the preceding discussion, has been quite useful for the examination of a number of small, non-volatile biomolecules [6.1].

6.3.1 Development of Matrix-Assisted Laser Desorption

While laser desorption mass spectrometry has been used to examine biomolecules with molecular weights up to 4000 amu, until recently only limited success had been achieved with larger biomolecules. A modified laser-desorption technique developed in 1987 [6.43] exploited the use of an absorbing matrix to soften the desorption/ionization of large biomolecules. This technique, dubbed "Matrix-Assisted Laser Desorption/Ionization" or MALDI has revolutionized the application of mass spectrometry for large biomolecules and has been successfully combined with TOF mass spectrometry for the detection of peptides with masses exceeding 300 000 amu, as summarized in a recent review article [6.3]. This MALDI technique, developed for TOF mass spectrometers, was initially demonstrated with a nicotinic acid matrix and 266 nm laser radiation from a pulsed Nd:YAG laser. For these experiments, a small amount of a peptide (typically a few picomoles) is mixed with a large excess of matrix compound such as nicotinic acid (a few microliters of a mM aqueous solution of matrix compound) which absorbs strongly at the laser wavelength. The resulting mixture is air-dried on a solid substrate. The laser radiation (typically low 10^6 W/cm^2) is primarily absorbed by the matrix compound rather than the biological analyte. The matrix "explodes" upon laser irradiation and the biomolecule is carried into the gas phase where it is ionized by proton transfer. For MALDI experiments, a "good" matrix compound, which produces primarily molecular ions with little or no fragmentation, must meet a number of criteria: it must strongly absorb the laser radiation, have similar solubility properties as the analyte, isolate the biomolecules from each other, co-crystallize with the biomolecule analyte, provide good ionization conditions (proton donation or abstraction), and not produce ionic interferences with the biomolecule. Since the initial development of MALDI, many other matrix/laser–wavelength combinations which are useful for studying large biomolecules have been reported.

In addition to its ability to produce ions from extremely large biomolecules, MALDI has a number of other features which make it especially applicable to the analysis of biological compounds. First, because MALDI spectra are simple and usually contain only molecular ions such as $(M + H)^+$ and $(M + 2H)^{2+}$ for each compound with virtually no fragment ions, this technique is quite suitable for the direct characterization of biological mixtures with minimal need for extensive sample preparation or chromatographic isolation. Second, unlike many other ionization methods for larger biomolecules, MALDI is generally found to be insensitive to contaminants, such as salts, buffers and surfactants, which are difficult to completely eliminate from a biological sample [6.3].

6.3.2 Interfacing MALDI with FTMS

The combination of MALDI with FTMS was initially demonstrated for the generation of molecular ions from small peptides [6.44, 45] and oligonucleotides [6.46, 47] and demonstrated that the presence of the matrix substantially reduced the amount of fragmentation of these biomolecules. This technique proved to be useful for determining the molecular weights, sequences and adduction sites for a number of small oligonucleotides at the low picomole level [6.46, 47]. The negative-ion MALDI-FTMS spectra for these biomolecules revealed abundant $(M - H)^-$ ions as well as fragment ions, as shown in Fig. 6.5 for the tetranucleotide d(AGCT), that provided the information necessary to

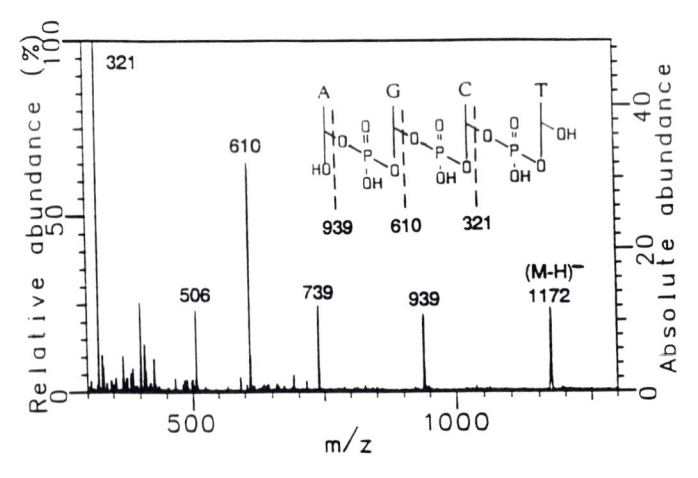

Fig. 6.5. The MALDI-FTMS negative-ion FTMS spectrum of d(AGCT) obtained with a 2-pyrazine carboxylic acid matrix and 266 nm laser radiation

determine oligonucleotide sequence and differentiate isomers. The primary fragmentation of these biomolecules was observed to be cleavage of the phosphate-ester bonds with the resulting charge retained on the 3' end of the oligomer, producing abundant fragment ions at m/z 939, 610 and 321 for d(AGCT). Fragmentation in the reverse direction with the charge retained on the 5' end was observed to be minor. This preferential fragmentation provides information which is sufficient to not only determine the sequence of an oligomer but also identify the 3' nucleotide of the oligomer and can be used to differentiate sequence isomers such as d(5'-CGCG-3') and (5'-CCGG-3') as well as isomers differing only in the location of the 3' end, such as d(5'-CGCG-3') and d(5'-GCGC-3') [6.46]. The CAD spectrum of the $(M - H)^-$ ion at m/z 1172 for d(AGCT) gave fragment ions identical to those generated by the laser-desorption process, indicating a preference for fragmentation of the phosphate-ester linkages. MALDI-FTMS was found to be useful for the structural characterization of modified oligonucleotides as well. For example, mass spectra for the tetranucleotide d(TGCA) were compared with mass spectra of a modified version of this tetramer, d(TG*CA) where G* is O^6-methyl guanine [6.46]. Inspection of the differences between the fragment ions for these similar biomolecules verified that the methyl adduct was present on the third nucleotide (guanine in this case) of the tetranucleotide.

6.3.3 Ion-Trapping Considerations for MALDI-FTMS

Initial investigations of MALDI-FTMS for larger molecules by several research groups were unsuccessful, in strong contrast to reports of spectra of proteins and peptides with molecular masses in excess of 100 000 amu obtained with MALDI-TOF. It became apparent that factors other than ion production needed to be addressed before MALDI could be successfully coupled with FTMS for the examination of larger biomolecules.

High-mass FTMS detection had previously been demonstrated for large cesium-iodide ionic clusters generated by Cs^+ Secondary Ion-Mass Spectrometry (SIMS) in a 7 T FTMS instrument [6.48]. In this case, ions with m/z exceeding 30 000 were trapped and detected by FTMS indicating that there were no inherent instrumental limitations for the FTMS detection of ions in this mass range. Based on these results, problems with efficiently trapping high-mass ions was thought to be the major factor limiting the success of the MALDI-FTMS experiments. The large cluster ions formed in a Cs^+-secondary ion source most likely have substantially different energetic characteristics from biopolymer ions generated by the MALDI process and may be trapped more efficiently in the FTMS cell. Indeed, TOF studies by several research groups have indicated that MALDI-generated ions are created with significant translational energies, and that these energies increase with increasing mass of the ions [6.49, 50]. Experimentally, ions with $m/z < 2000$, such as the matrix ions and small biomolecules, have kinetic energies in the low-eV range, while ions with $m/z > 5000$ have kinetic energies substantially larger than 10 eV. In the typical FTMS experiment, trapping voltages are maintained between 1 V and 4 V; higher voltages can induce substantial ion loss from the FTMS cell. Under these trapping potentials, these high-mass ions with high energies would not be efficiently trapped in the FTMS analyzer cell. This explanation supports early MALDI-FTMS results where spectra were obtained for biomolecules in the $m/z < 4000$ region, but not for larger compounds [6.44].

A variety of experimental approaches have been taken to attempt to improve the trapping efficiencies of FTMS for large ions. Initial experiments were directed at varying either the voltages applied to the laser probe tip [6.47] or the trapping plates [6.51] to more efficiently collect the high-mass ions with large kinetic energies. Although varying the probe-tip voltage did not provide better control of the MALDI ions, increasing the trapping plate voltages to 9 V did afford some improvement for biomolecules up to m/z 3000. Ejection of ions perpendicular to the magnetic field induced by these large trapping voltages was found to be a significant problem for ions with higher masses. Another approach to improve ion trapping involved high-pressure collisional cooling of the MALDI-generated ions to reduce the excess energy of the ions. This was accomplished by using a specially-designed high-pressure ion-source region for the laser desorption experiments [6.51]. While this approach was difficult to implement with the low-pressure requirements of the FTMS, it did provide some enhancement in the mass range for MALDI-generated ions, allowing the detection of bovine insulin (m/z 5734) and its dimer at $m/z > 10000$.

To date, the most successful approach for trapping high-mass ions involves electrostatic control of the MALDI-generated ions to remove excess energy and enhance trapping [6.52]. In this experiment, a special FTMS experimental sequence was used in which a 9 V differential was applied between the two trapping plates for times ranging from 100–200 µs after the laser pulse. This 9 V potential energy "hill" acted to electrostatically decelerate the MALDI-generated ions as they entered the FTMS cell. The trapping plates were then adjusted to normal values (3 V or less) for the remainder of the experiments. This approach proved to be successful for the detection of molecular ions from melittin, bovine insulin, cytochrome c, myoglobin and trypsinogen, whose molecular masses range from 2500 to 22000 amu. The positive ion spectrum for cytochrome c, shown in Fig. 6.6, reveals an abundant molecular ion with no fragmentation. (The ion at m/z 25000 is the proton-bound dimer of cytochrome c.) These experiments employed a 7 T magnet and 308 nm excimer-laser desorption/sinapinic acid MALDI conditions. While unit-mass resolution was obtained for melittin in these experiments, the resolutions of the larger ions were very poor. Attempts to improve resolution by increasing the ion-trapping time to permit laser-desorbed neutral molecules to be pumped away, thereby lowering the pressure for ion detection, were unsuccessful and resulted in substantial ion loss. This poor resolution was initially postulated to be due either to metastable decomposition of the MALDI ions or ion loss due to trapping inefficiencies (particularly in z-axis).

[Figure: low-resolution MALDI-FTMS spectrum with m/z axis from 5000 to 40000]

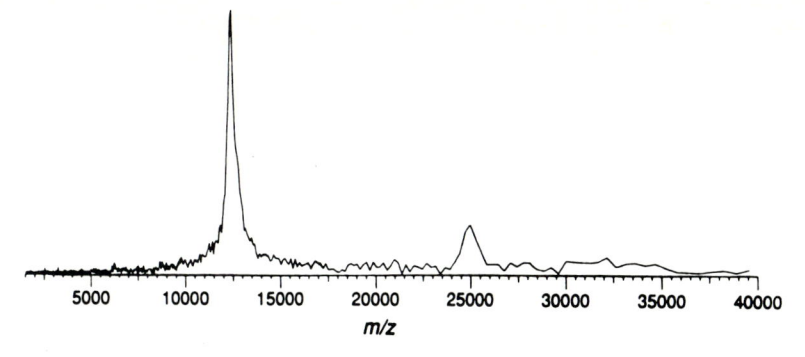

Fig. 6.6. The low-resolution MALDI-FTMS positive-ion FTMS spectrum of cytochrome c (MW 12 358.34) obtained with 308 nm excimer-laser radiation and a sinapinic acid matrix. Adapted from [6.52]

In order to improve the resolution of the MALDI-FTMS spectra, methods were examined to collisionally cool the MALDI ions in conjunction with the use of electrostatic deceleration techniques. It was found that adding sugars (such as sucrose or fructose) to the MALDI matrix substantially improved the mass resolution, as demonstrated for the molecular ion of bovine insulin at m/z 5733.69 shown in Fig. 6.7 [6.53]. The authors propose that the sugars decompose to water and carbon dioxide upon laser desorption, creating a momentarily dense gas environment for collisionally cooling the biomolecule ions.

[Figure: high-resolution MALDI-FTMS spectrum showing m/z 5733.69 with m/Δm = 11,580]

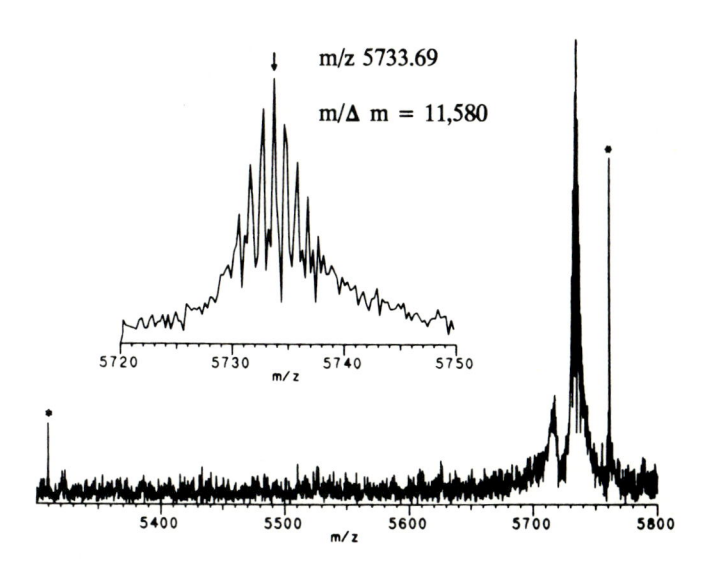

Fig. 6.7. The high-resolution MALDI-FTMS positive-ion FTMS spectrum of bovine insulin (MW 5733.69) obtained with 355 nm laser radiation and a 2,5-dihydroxybenzoic acid/d-ribose matrix (the starred peak is a frequency-noise spike from the instrument and was present even when the sample was not applied). Polyethylene glycol 1000 was used as a mass calibrant in a separate measurement. Adapted from [6.53]

6.3.4 Combining Separation Methods with MALDI-FTMS

To enhance its utility for the analysis of biological samples, MALDI-FTMS has been combined with off-line detection of chromatographically separated samples. For example, the MALDI-matrix compound can be mixed concurrently with biomolecules eluting from a capillary-zone electrophoresis column and the resulting mixture collected on a solid substrate, which is then analyzed by MALDI-FTMS [6.54]. MALDI-FTMS spectra were obtained at the low picomole level for somatostatin (MW 1 638), equine myoglobin (MW 16 951) and bovine insulin (MW 5 734) separated by this technique.

Slab-gel electrophoresis, another commonly used technique for the purification of biomolecules, has also been combined off-line with MALDI-FTMS for the detection of a range of oligonucleotides. Two methods of isolating the biomolecules from the gels, "freeze–squeeze" extraction for agarose gels [6.55] and electroelution for PAGE gels [6.56], were useful for extracting picomole levels of small oligonucleotides from gels, and should be applicable to larger biomolecules as well. For these studies, the MS/MS capabilities of FTMS were used to differentiate isomeric oligonucleotides extracted from the gels, indicating that the mass spectrometer might be useful for the identification of compounds which may be incompletely separated on the gel.

6.4 Future Directions

The combination of lasers with FTMS has provided a powerful instrumental method for the examination of a wide variety of compounds, ranging from clusters to biomolecules. The capability of this technique for the generation and examination of unusual species such as homonuclear and heteronuclear clusters as well as the "soft ionization" of large non-volatile biomolecules has had tremendous impact on the advancement of these areas of research.

Work is in progress in a number of laboratories to investigate and extend the practical upper-mass limit and mass resolution achievable for large ions by FTMS. The demonstration of laser-ablation FTMS for large cluster ions and MALDI-FTMS for biomolecules with MW > 20 000 has prompted renewed interest in probing the instrumental factors which influence the achievable mass range. In the FTMS experiment, increasing the magnetic field strength, as discussed previously, will profoundly enhance both upper-mass range and mass resolution. For example, a theoretical investigation of the factors which control ion trapping concluded that the upper-mass limit for *thermal* ions would be m/z 50 000 in a 3 T magnet, m/z 300 000 in a 7 T magnet and m/z > 1 000 000 for a 14 T magnet [6.2]. However, recent investigations have shown that mass limits based solely on field strength may not completely reflect all the experimental factors affecting high-mass measurement. As pointed out earlier, MALDI ions are known to have non-thermal kinetic energy distributions, and research will continue to address better methods for controlling and cooling the MALDI ions in order to extend the effective mass range for MALDI-FTMS. The behavior of low-mass ions in low magnetic fields have been examined, in hopes of scaling observations to large ions in high magnetic fields [6.57]. These experiments suggest that correction of electronic-field effects by new cell designs can address resolution and mass-range limitations.

A variety of different cell geometries have been tested to improve the electrostatic fields in the cell for better ion trapping and detection of high-mass ions. One promising new FTMS cell is a modified cubic cell in which screens have been placed just inside the trapping plates [6.4, 5]. By grounding the screens while simultaneously applying an electrostatic potential to the trapping plates, the electric field can be made uniformly flat across the entire FTMS cell. This cell design has been demonstrated to improve trapping efficiencies as well as resolution and mass-measurement accuracy.

In addition to instrumentation development, current research in FTMS is also directed at the development of techniques for obtaining structural information on large ions. The traditional ion-structural techniques that have been successfully used for investigation of smaller ions, such as CAD and photo-induced dissociation, may not be as effective for large, singly charged ions. In fact, CAD has been found to be fairly limited for the structural characterization of ions with m/z exceeding 3 000 because the center of mass energy for a collision of a large ion with a small neutral atom, such as argon, is small. Similar limitations would be expected for photo-induced fragmentation experiments. A potential alternative for structural characterization of large molecules is Surface-Induced Dissociation (SID) [6.58]. This experiment is accomplished by colliding an ion with a surface, and results in much higher energy deposition into the ion than possible with CAD, which might make it more suitable for the dissociation of large ions. SID has been demonstrated with FTMS [6.59, 60], although the maximum energy deposition and amount of neutralization which occurs is unknown at this early stage of development. Selective ion–molecule reactions between large ions and suitable reagents may also provide a means of probing structures. For example, change-exchange reactions will continue to be useful for probing the electronic properties of these ions. In addition, reactive reagents might be useful for performing selective cleavages in a biomolecule in a similar manner to enzymatic reactions which are used in biochemical procedures.

The future should also bring additional improvements in the hardware associated with the FTMS, such as electronics, computers and superconducting magnets, which will lead to enhanced instrument performance. Magnets with field strengths of 7 T are already being employed to extend mass range and enhance mass resolution, and experiments with even higher field magnets (12–20 T) are currently in the planning stage. Improvements in excitation and detection electronics and computer technology for data manipulation are already beginning to impact the performance of FTMS. Future advances in these technologies will also undoubtedly enhance the instrumental capabilities of FTMS for more sensitive ion detection, better resolution and higher mass accuracies.

6.5 Conclusions

Although several different mass analyzers are compatible with pulsed lasers, Fourier transform mass spectrometry offers a number of additional capabilities of probing ion structures in detail. The high-resolution and accurate mass measurement, ion-trapping and ion-manipulation capabilities of FTMS make this a powerful technique for examining laser-generated or laser-desorbed ions. In particular, collision-activated dissociation, ion-molecule reactions and photodissociation have been used to examine the structures and reactivities of ions ranging from metal clusters and fullerenes to biomolecules.

Acknowledgements. Acknowledgement is given to Michelle Buchanan, Greg Hurst, Elizabeth Stemmler and Marc Wise for technical comments on this manuscript, and to the Department of Energy, Office of Health and Environmental Research for support of FTMS research on biomolecules under contract DE-AC05-84OR21400 with Martin Marietta Energy Systems, Inc.

References

6.1 D. Lubman (ed.): *Lasers and Mass Spectrometry* (Oxford Univ. Press, New York 1990)
6.2 M.A. May, P.B. Grosshans, A.G. Marshall: Int. J. Mass. Spectrom. Proc. **120**, 193 (1991)
6.3 F. Hillenkamp, M. Karas, R.C. Beavis, B.T. Chait: Anal. Chem. **63**, 1193A (1991)
6.4 A.G. Marshall, L. Schweikhard: Int. J Mass Spectrom. Ion Proc. **118/119**, 37 (1992)
6.5 A.G. Marshall, P.B. Grosshans: Anal. Chem. **63**, 215A (1991)
6.6 E.R. Williams, K.D. Henry, F.W. McLafferty: J. Am. Chem. Soc. **112**, 6157 (1990)
6.7 F.W. McLafferty (ed.): *Tandem Mass Spectrometry* (Wiley, New York 1983)
6.8 E.R. Williams, F.W. McLafferty: J. Am. Soc. Mass Spectrom. **1**, 361 (1990).
6.9 P. Jena, B.K. Rao, S.N. Khanna (eds.): *Physics and Chemistry of Small Clusters* (Plenum, New York 1987)
6.10 E.L. Muetterties: Bull. Soc. Chem. Belg. **84**, 959 (1975)
6.11 E. Shustorovich, R.C. Baetzold, E.L. Muetterties: J. Phys. Chem. **876**, 1100 (1983)
6.12 R.E. Smalley: In *Comparison of Ab-Initio Quantum Chemistry with Experiment for Small Molecules: State of the Art*, ed. by R.J. Bartlett (Reidel, New York 1985) p. 53
6.13 R.E. Smalley: In *Metal-Metal Bonds and Clusters in Chemistry and Catalysis*, ed. by J.P. Frackler (Plenum, New York 1989) p. 249
6.14 R.P. Andres, R.S. Averback, W.L. Brown, L.E. Brus, W.A. Goddard, A. Kaldor, S.G. Louie, M. Moscovits, P.S. Peercy, S.J. Riley, R.W. Siegel, F. Spaepen, Y. Wang: J. Mater. Res. **4**, 704 (1989)
6.15 M.E. Tremblay, B.W. Smith, M.B. Long, J.D. Winefordner: Spectrosc. Lett. **20**, 311 (1987)
6.16 M.L. Mandich, W.D. Reents Jr., V.E. Bondybey: J. Phys. Chem. **90**, 2315 (1986)
6.17 W.D. Reents Jr., M.L. Mandich, V.E. Bondybey: Chem. Phys. Lett. **125**, 324 (1986)
6.18 W.D. Reents Jr., M.L. Mandich, V.E. Bondybey: Chem. Phys. Lett. **131**, 1 (1986)

6.19 A. O'Keefe, M.M. Ross, A.P. Baronavski: Chem. Phys. Lett. **130**, 17 (1986)
6.20 S.W. McElvany, W.R. Creasy, A.O. O'Keefe: J. Chem. Phys. **85**, 632 (1986)
6.21 S.W. McElvany, H.H. Nelson, A.P. Baronovski, P. Andrew, C. H. Clifford, J.R. Eyler: Chem. Phys. Lett. **134**, 214 (1987)
6.22 W.R. Creasy, J.T. Brenna: Chem. Phys. **126**, 453 (1988)
6.23 R.L. Hettich: J. Am. Chem. Soc. **111**, 8582 (1989)
6.24 T.G. Dietz, M.A. Duncan, D.E. Powers, R.E. Smalley: J. Chem. Phys. **74**, 6511 (1981)
6.25 D.E. Powers, S.G. Hanson, M.E. Geusic, A.C. Puiu, J.B. Hopkins, T.G. Dietz, M.A. Duncan, P.R.R. Langridge-Smith, R.E. Smalley: J. Phys. Chem. **86**, 2556 (1982).
6.26 S.J. Riley, E.K. Parks, C.R. Mao, L.G. Pobo, S. Wexler: J. Phys. Chem. **86**, 3911 (1982)
6.27 V.E. Bondybey, J.H. English: J. Chem. Phys. **80**, 568 (1984)
6.28 J.M. Alford, F.K. Weiss, R.T. Laaksonnen, R.E. Smalley: J. Phys. Chem. **90**, 4480 (1986)
6.29 S. Maruyama, L.R. Anderson, R.E. Smalley: Rev. Sci. Instrum. **61**, 3686 (1990)
6.30 D.B. Jacobson, B.S. Freiser: J. Am. Chem. Soc. **106**, 5351 (1984)
6.31 R.L. Hettich, B.S. Freiser: In *Fourier Transform Mass Spectrometry: Evolution, Innovation, and Applications*, ed. by M.V. Buchanan, Vol. 359 (Am. Chem. Soc. Washington, DC 1987) p. 155
6.32 D.B. Jacobson, B.S. Freiser: J. Am. Chem. Soc. **108**, 27 (1986)
6.33 R.L. Hettich, B.S. Freiser: J. Am. Chem. Soc. **109**, 3537 (1987)
6.34 R.L. Hettich, B.S. Freiser: J. Am. Chem. Soc. **107**, 6222 (1985)
6.35 R.L. Hettich, T.C. Jackson, E.M. Stanko, B.S. Freiser: J. Am. Chem. Soc. **108**, 5086 (1986)
6.36 P.R. Buseck, S.J. Tsipursky, R.L. Hettich: Science **257**, 215 (1992)
6.37 T. Guo, C. Jin, R.E. Smalley: J. Phys. Chem. **95**, 4948 (1991)
6.38 R.L. Hettich, R.N. Compton, R.H. Ritchie: Phys. Rev. Lett. **67**, 1242 (1991)
6.39 P.A. Limbach, L. Schweikhard, K.A. Cowen, M.T. McDermott, A.G. Marshall, J.V. Coe: J. Am. Chem. Soc. **113**, 6795 (1991)
6.40 J.A. Zimmerman, J.R. Fyler, S.B.H. Bach, S.W. McElvany: J. Chem. Phys. **94**, 3556 (1991)
6.41 R.T. Laaksonen: Hydrogen chemisorption on transition metal clusters. Ph.D. Thesis, Rice University, Houston, TX (1991)
6.42 M.A. Nygren, P.E.M. Siegbahn, C. Jin, T. Guo, R.E. Smalley: J. Chem. Phys. **95**, 6181 (1991)
6.43 M. Karas, K. Bachmann, U. Bahr, F. Hillenkamp: Int J. Mass Spectrom. Ion Proc. **78**, 53 (1987)
6.44 R.L. Hettich, M.V. Buchanan: J. Am. Soc. Mass Spectrom. **2**, 22 (1991)
6.45 L.M. Nuwaysir, C.L. Wilkins: SPIE **1437**, 112 (1991)
6.46 R.L. Hettich, M.V. Buchanan: J. Am. Soc. Mass Spectrom. **2**, 402 (1991)
6.47 R.L. Hettich, M.V. Buchanan: Int. J. Mass Spectrom. Ion Phys. **111**, 365 (1991)
6.48 C.B. Lebrilla, D.T.-S. Wang, R.L. Hunter, R.T. McIver, Jr.: Anal. Chem. **62**, 878 (1990)
6.49 R.C. Beavis, B.T. Chait: Chem. Phys. Lett. **181**, 479 (1991)
6.50 Y. Pan, R.J. Cotter: Org. Mass Spectrom. **27**, 3 (1992)
6.51 T. Solouki, D.H. Russell: Proc. Natl. Acad. Sci. USA **89**, 5701 (1992)
6.52 J.A. Castoro, C. Koster, C. Wilkins: Rapid Comm. Mass Spectrom. **6**, 239 (1992)
6.53 C. Koster, J.A. Castoro, C.L. Wilkins: J. Am. Chem. Soc. **114**, 7572 (1992)
6.54 J.A. Castoro, R.W. Chiu, C.A. Monnig, C.L. Wilkins: J. Am. Chem. Soc. **114**, 7571 (1992)
6.55 J.C. Dunphy, K.L. Busch, R.L. Hettich, M.V. Buchanan: Anal. Chem. **65**, 1329 (1993)
6.56 J.C. Dunphy, K.L. Busch, R.L. Hettich, M.V. Buchanan: unpublished results
6.57 D.L. Rempel, R.P. Grese, M.L. Gross: Int. J. Mass. Spectrom. Ion Proc. **100**, 381 (1990)
6.58 M.D.A. Mabud, M.J. Dekrey, R.G. Cooks: Int. J. Mass. Spectrom. Ion Proc. **67**, 285 (1985)
6.59 E.R. Williams, K.D. Henry, F.W. McLafferty, J. Shabanowitz, D.F. Hunt: J. Am. Soc. Mass Spectrom. **1**, 413 (1990)
6.60 J.A. Castoro, L.M. Nuwaysir, C.F. Ijames, C.L. Wilkins: Anal. Chem. **64**, 2238 (1992)

7. Diagnostic Studies of Laser Ablation for Chemical Analysis

A.D. Sappey and *N.S. Nogar*

With 5 Figures

Ever since the first report of laser action, it has been recognized that the study of laser–material interactions provides information on a wealth of extremely interesting physics and chemistry, and also comprises a practical tool for many applications [7.1]. Many physical regimes have been examined from low-irradiance thermal evaporation through the regime of microplasma formation to the intensities associated with inertial confinement fusion. The physical processes associated with these interactions also span a wide range, from simple linear absorption with associated small temperature fluctuations up through highly nonlinear interactions, such as multiphoton absorption and inverse Bremsstrahlung with explosive-like compaction and expansion of the irradiated material.

Similarly, the uses of laser ablation are extremely numerous. These include both practical applications [7.2] such as thin- and thick-film growth [7.3], microparticle production, surface cleaning [7.4], etching and patterning [7.5], intertial confinement fusion and sampling for chemical analysis [7.6], as well as more fundamental pursuits, such as surface-kinetic studies [7.7], optical-damage investigations [7.8] and detailed studies of laser-material interactions.

Our interest here is primarily in the use of laser sampling for chemical analysis. Laser ablation has been used, and is being considered for use, in a wide range of field-diagnostic applications. In part, this development is being driven by the recent emphasis on environmental monitoring and remediation. For example, currently available technology for characterization of soil samples from a potentially contaminated site is expensive and slow. It involves careful remote soil sampling, sample preparation and analysis by, perhaps, several techniques. Normally only the sampling occurs "on site"; the sample preparation and analysis are performed in a laboratory some distance away. As a result, the analysis may be too slow to prevent further environmental insult. The application of laser-based analytical techniques may eventually result in real-time, on-site sample characterization.

Several laser-ablation based techniques show great promise. One such technique, Laser-Induced Breakdown Spectroscopy (LIBS), potentially circumvents all of the problems with the current technologies [7.9]. A LIBS experiment consists of focussing a high power, pulsed laser onto a sample in the ambient atmosphere, thereby creating a microplasma. Light from the microplasma is dispersed in a small monochromator to detect the characteristic emission from various atomic and molecular species such as heavy metals which are present in the soil. Clearly, details of the observed spectra will depend on the temporal and spatial evolution of the plasma, which, in turn, depends on the atmosphere (or lack thereof) in which the ablation occurs. An instrument for the quantitative measurement of beryllium in contaminated soil samples based on the LIBS technique has already been designed and demonstrated at Los Alamos. The unit is the size of a small suitcase and operates on battery power making it ideal for remote-site characterization.

A more species-selective variant of LIBS involves probing the plasma with a second laser beam to perform Laser-Induced Fluorescence, LIF [7.10]. This technique is more complicated requiring a second, tunable light source. However, it has a decided advantage in selectivity since the species of interest can be selected by both excitation and fluorescence wavelengths. This may be necessary for detection of species which emit at wavelengths too close to resolve using the more simple LIBS apparatus.

Another application of laser ablation is being developed for atomic emission and mass-spectral-analysis of Inductively Coupled Plasmas (ICP's) [7.11]. It is presently extremely cumbersome to introduce solid samples into an inductively coupled plasma spectrometer. Nominally, this involves reducing the solid to a solution that is nebulized into the plasma. However, it has recently been shown that laser ablation of solids can be used to introduce samples directly into the torch with the feed gas stream. A further application already demonstrated to be useful is laser ablation/mass spectrometry [7.12]. This is a very powerful technique, but is normally applied only under relatively high-vacuum conditions.

In the remainder of this chapter we focus primarily on the applications of laser ablation for chemical analysis [7.2, 6]. This will include both ablation into vacuum, where the principal method of analysis is mass spectral, and evaporation into an ambient atmosphere, where primarily optical diagnostics are employed. Since both of these applications depend, in some detail, on the mechanisms and processes of laser ablation, including particularly plume evolution, much of this chapter will be devoted to diagnostic methods and the application of these results to the effective use of lasers in sampling.

7.1 Laser Ablation in Vacuum

The most common analytical application for laser sampling into a vacuum is mass spectral analysis [7.13]. Since its original description, numerous research papers and review articles [7.12, 14–18] have appeared on various aspects of laser mass spectrometry. Pulsed lasers are typically used in these experiments, and sample vaporization and ionization generally takes place in one of three modes. In one case, relatively high intensity pulses. $I \geq 100\,\mathrm{MW/cm^2}$, are used to simultaneously vaporize, ionize, and in some cases fragment the sample of interest [7.17]. This type of instrumentation is often referred to as a laser microprobe, or laser ionization mass spectrometer [7.18-20]. It is available from several vendors as a commercial instrument. In the second case, relatively gentle irradiation, $I \leq 10 \text{ MW/cm}^2$, is used to vaporize intact or largely intact molecules. In this type of instrumentation, ionization and/or fragmentation occurs by a secondary process, typically electron impact, chemical ionization, plasma, or photoionization by a second laser [7.21], subsequent to volatilization. The third method, known as Matrix-Assisted Laser Desorption/Ionization (MALDI), is a hybrid of these types, normally applied to the analysis of large molecules. In this method, the sample, which is contained in a special sample matrix [7.22, 23], is subjected to low energy, pulsed UV-laser irradiation. The presence of the matrix causes desorption of large intact molecular or quasi-molecular ions.

Laser sampling exhibits a number of useful properties for chemical analysis [7.24]. 1) No background is introduced due to bulk heating of the sample. 2) Spatial resolution can be very good, limited only by diffraction of the incident beam (typically ≈ 1 μm in diameter) and, at high fluxes, cratering of the surface (typically ≈ 10 μm). This capability can be very important for small samples, and for samples containing inhomogeneously distributed components. 3) Little sample preparation is needed. While conventional wet analytical samples can be dried on a sample introduction device, non-conventional samples can also be introduced directly into the source region of the mass spectrometer. 4) Sensitivity is excellent, and the detection limit frequently falls in the femtogram to attogram (absolute) or sub-ppb range. 5) The possibility exists for absolute measurements without recourse to standard samples.

7.1.1 Instrumentation

A variety of laser sources have been used in laser evaporation for mass spectrometry. In the vast majority of experiments, pulsed lasers have been used. Q-switched ruby lasers were used in many early experiments [7.25]. These instruments produce large pulse energies ($E_p \geq 1$ J), but at extremely low repetition rates, $f \leq 1$ Hz. This latter constraint is a distinct disadvantage for a process such as laser evaporation, which is highly stochastic, and for which extensive signal averaging may be required for quantitative results.

More frequently, TEA CO_2 lasers [7.26], Q-switched Nd^{3+}: YAG [7.21], nitrogen lasers [7.27], and rare gas/halogen excimer lasers [7.28] are used for laser evaporation. All can operate at repetition rates $f \geq 10$ Hz, produce pulse energies $E_p \geq 100$ mJ (with the exception of nitrogen lasers for which $E_p \leq 10$ mJ), and can produce relatively short pulses, ≤ 100 ns. The wavelength of operation often controls the mode of evaporation. Long wavelength (≈ 10 μm) CO_2 laser pulses usually induce a "thermal" evaporation process, in which the evaporated material exhibits well-defined translational and internal temperatures in equilibrium with the local surface temperature [7.29]. Commercial CO_2 lasers are relatively inexpensive and reliable. Nd^{3+}: YAG lasers offer a more expensive, but also more versatile tool for laser evaporation studies. Frequency multiplication of the output from a Nd^{3+}: YAG laser allows a number of wavelengths to be accessed. Operation at the fundamental output wavelength (1.06 μm) usually results in thermal evaporation processes, similar to those observed with CO_2-laser initiation. Operation at the fourth harmonic (266 nm) or a Raman-shifted variant [7.30], on the other hand, often initiates processes which are non-thermal in nature. The result may be non-thermal velocity or internal energy distributions, and/or unusual fragmentation patterns. Intermediate wavelengths, 532 nm and 355 nm, frequently produce ambiguous results which can be characterized as neither thermal nor photophysical in nature [7.8]. Excimer lasers are normally operated at 308 nm (XeCl), 248 nm (KrF) or 193 nm (ArF), and may produce a number of novel and unusual results at short wavelengths [7.31]. In addition, in instances where a micro-plasma is produced, the wavelength of irradiation determines the fractional absorption of the beam. The plasma resonant frequency v_p is given by [7.16]:

$$v_p = (4\pi n_e e^2/m_e)^{1/2}$$
$$= 8.9 \times 10^3 (n_e)^{1/2} , \quad (7.1)$$

where n_e is the electron number density (cm^{-3}) in the plasma, e is the electron charge, and m_e is the electron mass. In order for substantial absorption to occur in the plasma, $v_1/v_p \approx 1$, where v_1 is the laser frequency. Otherwise, the laser radiation is reflected.

Similarly, the amount of material removed from the surface under conditions of microplasma formation is given by [7.16]:

$$m = 110 \lambda^{-4/3} \left(\frac{\phi_a}{10^{14}} \right)^{1/3} \quad (7.2)$$

where m is the mass ablation rate (kg/s$^1 \cdot$cm^2) and ϕ is the absorbed flux (W/cm^2), and λ is the wavelength in μm.

Irradiation of the sample may occur either from the "backside" (irradiated surface opposite to the mass spectrometer flight tube), or from the "frontside" (laser irradiation falling on the side exposed to the flight tube). In the former case, thin samples and supports are required, while in the latter case more latitude is allowed in sample characteristics. An alignment laser and microscope to view the sample surface are included in many instruments. It is often advantageous to have a mechanism for rastering the sample in front of the laser beam to allow either for signal averaging in homogeneous samples, or to measure the distribution of material in inhomogeneous samples [7.14].

For experiments in which lasers are also used for post-ionization, the common instrument of choice is a pulsed tunable laser, often with wavelength extension capabilities [7.32] such as frequency doubling, frequency mixing or Raman shifting. The tuning capability allows the use of resonant intermediate states in a multistep ionization process, thus increasing both the probability of ionization for most atoms [7.33] and molecules and the selectivity of ionization. The latter capability can be important in analyzing complex mixtures [7.34]. In some cases a flashlamp-pumped laser is used, but a wider tuning range and shorter pulses are usually obtained with Nd^{3+}: YAG or XeCl-excimer laser pumping. The tuning range is typically from the near IR ($\approx$ 800 nm) to the VUV ($\approx$ 200 nm), with 6–15 ns pulses from 1–20 mJ, and repetition rates of 10–500 Hz. With the use of resonant intermediate states, the ionization can usually be saturated within the laser focal volume ($\approx$ 100% conversion of neutrals to ions).

In some instances, a high-power fixed-frequency laser, typically a frequency quadrupled Nd^{3+}: YAG [7.35] or KrF-excimer laser [7.36], may be used for post-ablative ionization. Nonresonant ionization is usually somewhat less efficient than for resonant processes (typically 1–10%), although accidental coincidences with autoionizing transitions may increase the ionization probability by orders of magnitude [7.37]. On the other hand, the nonresonant process tends to ionize all species present in the desorbed plume, and so yields a broader range of information.

Only a small fraction of the published work in laser mass spectrometry has been performed with single-channel detection instruments, such as quadrupole and magnetic sector spectrometers. Magnetic instruments are typically used when superior sensitivity or large dynamic range is required. By far the majority of experiments have utilized multichannel detection, such as is possible in Time-Of-Flight (TOF), and more recently, Fourier Transform Ion Cyclotron Resonance spectrometers (FTICR). In fact, current interest in TOF instruments can be largely attributed to laser-based applications. Recent developments have resulted in conventional design TOF instruments of unprecedented resolving power, limited only by the temporal pulsewidth of the ionization laser. In addition, the development of reflecting TOF instruments has resulted in resolving powers $\geq 10^4$. A channel electron multiplier, venetian-blind type multiplier, channel plate or "Daly" detector is typically used for ion detection, while a transient digitizer is usually used to record the data.

FTICR mass spectrometers are extremely well-suited for diagnosis of the composition of transient vapor pulses [7.38]. FTMS offers the advantage of multiplex detection, superior resolving power (often $\geq 10^5$), excellent sensitivity (sub-pg, absolute) and very high ultimate mass range. The major drawback is the limited dynamic range, less than 300, a result both of space-charge restraints on the number of ions that can be stored in the ion trap, and on the method of detection, which precludes the use of high-dynamic range electron multiplier tubes. More recently, quadrupole ion trap mass spectrometers have also been coupled with lasers for both elemental and molecular analysis [7.39].

7.1.2 Physical Processes for Laser Ablation In Vacuo

The use of laser ablation for mass spectrometry requires some knowledge of the characteristics of the desorbed plume. In particular, it is useful to characterize the spatial and temporal evolution of the plumes' number density, state of ionization and chemical speciation. In the simplest case, where the plume is assumed to consist of discrete, neutral, monoatomic species, modeling calculations can be of use in evaluating the plume evolution, provided some simplifying assumptions are made [7.40]. These include:

— A point source for emission.
— The transient associated with material ejection is short compared to plume evolution and development.
— The plume is interrogated by a second laser beam at a constant distance from the sample surface (see Fig. 7.1).

We begin by calculating the flux density of atoms (or molecules) $\Gamma(v, R, t)$, with velocities in the range (v to $v + dv$) per unit area per unit time from the surface:

$$\Gamma(v, R, t_0) = v n_0 f(v, t_0) \cos(\theta) dv d\Theta , \qquad (7.3)$$

where θ is defined as the polar angle with respect to the surface normal and $n_0 f(v, t_0)$ is the distribution of molecules per unit volume in the appropriate velocity range at the time of evolution from the surface, and we have implicitly assumed an isotropic distribution (Fig. 7.1 shows both an isotropic distribution and a forward-peaked distribution characteristic of high-density ablation). Next,

Laser Desorption/Postionization

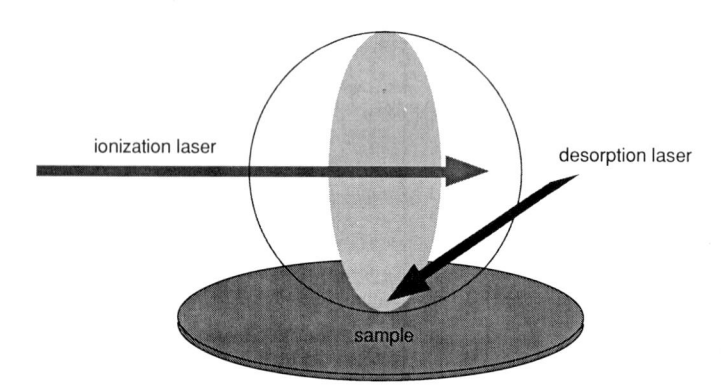

Fig. 7.1. Depicts a typical geometry for laser-ablation experiments. The open figure depicts a $\cos \Theta$ distribution, typical of low-density ablation, while the filled figure depicts a forward directed, $\cos^n \Theta$ distribution typical of high-density plumes. An ionization laser is depicted here, though a variety of probes can be used

it is convenient to introduce the speed distribution in terms of spherical polar coordinates $dv = v^2 \sin(\theta) d\theta\, d\phi\, dv$ and the solid angle element $d\Omega = \sin(\theta)\, d\theta\, d\phi$ to yield:

$$\Gamma(v, R) = n_0 v^3 f(v) \cos(\theta)\, d\Omega\, dv . \tag{7.4}$$

In general, the distribution will be integrated over a constant set of spatial coordinates, so that (7.4) may be written as:

$$\Gamma(v, R) = C v^3 f(v)\, dv , \tag{7.5}$$

where C includes the initial number density and the angular integration. Since the plume is interrogated by the second laser beam at a constant linear distance from the sample surface, $v = L/t$ and $dv = L\, dt/t^2$. If we assume a Maxwellian velocity distribution, then the signal generated (S) can be given by

$$S(t) = K t^{-4} \exp[-(L/t)^2/\sigma^2]\, dt , \tag{7.6}$$

where K includes terms in the spatial coordinate (distance from sample to interrogating laser beam) and the signal generation and collection efficiency, and σ is the local most-probable velocity $\sqrt{2kT/m}$. Another commonly observed velocity distribution is a shifted Maxwellian velocity distribution with an additional parameter v_0, to account for the center-of-mass velocity

$$S(t) = K t^{-4} \exp\{-[(L/t) - v_0]^2/\sigma^2\} . \tag{7.7}$$

This form of distribution has been shown previously to result from the adiabatic free-jet expansion [P.41, 42]. Laser ablation may also produce a distribution characteristic of a Knudsen layer, as modeled recently by *Kelly*, and *Dreyfus* and *Kelly* [P. 42, 43]. This model considers the formation of a Knudsen layer (when particles are no longer adequately described by a Maxwellian velocity distribution) via sputtered particle collisions and parallels shifted Maxwellian distributions. Here, as a result of collisions, ablated species acquire a common center-of-mass velocity rather than a common temperature. Using the formalism of *Kelly* and *Dreyfus*, the TOF signal is described by

$$S(t) = K t^{-4} \exp[-\beta_K^2 (L - u_K t)^{2/t^2}] . \tag{7.8}$$

In this model, u_K represents the center-of-mass velocity, while β_K is defined in terms of the Knudsen layer temperature T_K, $\beta_K = (m/2kT_k)^{1/2}$. The desorbing surface temperature is related to the Knudsen layer temperature through the heat-capacity ratio of the material in the plume.

All of these distributions have been observed experimentally [7.44], in some cases several in the same experiment [7.8]. Figure 7.2 shows the velocity distributions of calcium atoms desorbed from a calcium fluoride lattice. The Ca^+ observed in the laser-ionization experiments is produced from atomic Ca by a "2 + 1" (photons to resonance + photons to ionize) process through the

Ca FROM CaF₂ 1.06 μm

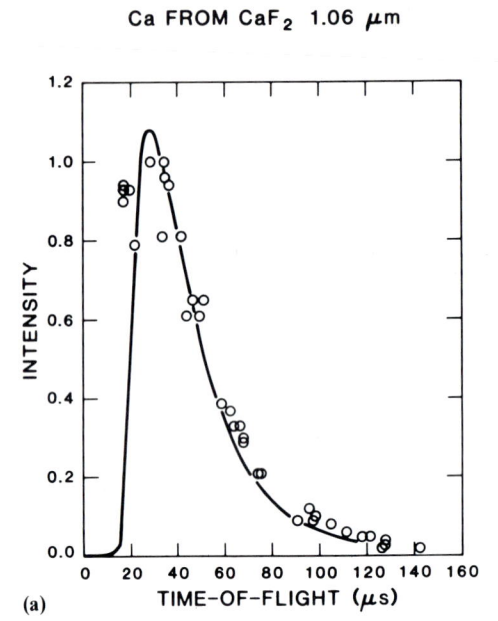

Fig. 7.2. Depicts arrival-time (velocity) distributions for Ca ablated from a CaF$_2$ crystal. The open circles are data, while the solid lines are fits to analytical distributions. Ablation initiated by 1.06 μm irradiation (**a**) exhibits a thermal distribution, while irradiation at 266 nm (**b**) generates a distribution best described as a sum of thermal and nonthermal distributions. In (**b**), the left-hand side shows two separate distributions, while the right-hand side shows the sum of two distributions

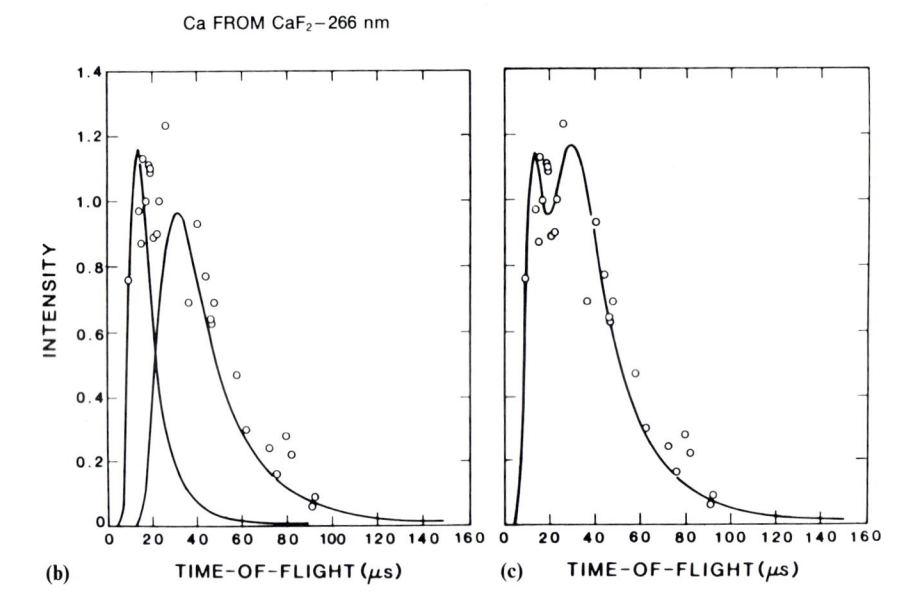

Ca FROM CaF₂ – 266 nm

enhancing $^1P_00$ state at 37298 cm^{-1} (CaF was also observed by an MPI process). Velocity distributions were measured by varying the time delay between the damage and ionization (probe) lasers. Ablation was initiated at the fundamental (1.06 μm) and frequency tripled (355 nm) and quadrupled (266 nm) wavelengths. The results were found to be dependent on wavelength and may be summarized as follows.

i) For 1.06 µm irradiation, we observed thermal (850 K) velocity distributions for both Ca and CaF. In addition, the CaF radical exhibited significant amounts of internal (rotational and vibrational) excitation, consistent with an internal "temperature" of $\approx 10^3$ K.
ii) For both 355 nm and 266 nm irradiation the velocity distributions were bimodal with a fraction ($\leq 50\%$) of the spalled material exhibiting very high (4000 K) kinetic temperatures, while the remainder exhibited a temperature similar (800–1000 K) to that observed for the 1.06 µm experiments. In addition, both the vibrational and rotational temperatures of the CaF radicals decreased with decreasing ablation wavelength.

The kinetic temperatures and CaF internal energies observed at 1.06 µm argue for a thermal or sonic mechanism with the evaporation of a relatively low-density plume. The results at shorter wavelengths suggest a more direct photo-physical interaction, where rapid energy deposition is followed by non-adiabatic transitions of the fragments to antibonding states resulting in the production of a relatively high-density plume. This, in turn, initiates a variety of collisional processes which can lead to nonthermal velocity distributions. Nonthermal velocity distributions are now recognized as a rather common occurrence in laser ablation both in vacuum and in an atmosphere.

7.1.3 Examples

Three examples will suffice to demonstrate the wide variety of applications for laser ablation in a vacuum: one in inorganic materials analysis, one in biochemical analysis, and one in geochemical analysis. In the first instance [7.44], laser post-ionization was used to measure speciation and velocity distribution of material ablated from a $YBa_2Cu_3O_{7-\delta}$ target. Laser ablation of bulk High-Temperature Superconducting (HTS) material provides a useful means of producing high-quality HTS thin films, and as such has been the object of much study. Mass-spectrometric probes of the ablation plume provided a microscopic understanding of the ablation event and plume development as well as providing a process monitor for thin-film production. Resonance Ionization Mass Spectrometry (RIMS) detection of the ablated (1.06 µm, 1 J/cm^2) neutral species supplied valuable physical information about the ablation event necessary for developing models of the process. TOF/RIMS allowed detection of Cu, Cu$_2$, Y, YO, Ba, and BaO ablated from a $YBa_2Cu_3O_{7-\delta}$ target. Approximately equal velocities were observed for all neutral species at constant ablation laser fluence. In addition, BaO exhibited a rotational temperature far less than the directed translational temperature.

In the biochemical arena, laser-ablation/RIMS has been proposed as a tool for DNA sequencing [7.45, 46]. Using either sputter-initiated RIMS or laser-ablation RIMS, it is possible to localize and quantify with approximately 1 µm spatial resolution ultratrace concentrations of a selected element at or near the surface of solid samples. The sensitivity and selectivity of the RIMS process is especially valuable for trace-element analysis in biological media, where the complexity of the matrix is frequently a serious source of interference. Both the sputter-initiated and laser-atomization techniques were compared to determine their characteristics to localize and quantify tin-Sn-labeled DNA.

A method was described for synthesis of a tin reagent, TriEthylStannyl-PropanoicAcid (TESPA) and its attachment to oligonucleotide primers. Except for the expected mobility retardation, the presence of ^{116}Sn-TESPA did not affect the sequencing ladder on electrophoresis gels. DNA bands on an electrophoresis gel were first located on an autoradiograph and then analyzed by resonance ionization spectroscopy to demonstrate coincidence of the signals.

Data were presented showing: (a) differences between sputter-initiated and laser-atomization response as a function of atomization parameter, substrate and analyte; and (b) the detection and resolution (spatial and isotopic) of sub-attomole quantities of Sn-labeled DNA bands. Both techniques have the potential of making a strong contribution to DNA sequencing.

In the last instance, laser ablation/laser ionization was used in a microprobe mode to analyze carbonaceous material from meteorites [7.47–49]. This analysis is nondestructive from a mineralogical point-of-view, and requires only milligram quantities of meteoritic material. In initial experiments, Polycyclic Aromatic Hydrocarbons (PAHs) in C1, C2, and C3 carbonaceous chondrites and in some ordinary chondrites were directly analyzed. The method, two-step laser-desorption/laser-ionization mass spectrometry, uses an IR laser to volatilize constituent molecules intact and a UV laser to ionize the desorbed molecules in a selective manner with little or no fragmentation. At the ionization wavelength of 266 nm, parent ion peaks of the PAHs dominate the mass spectra. Quantitative analysis is possible; as an example, the concentration of phenanthrene in the Murchison meteorite was 5.0 ppm. In addition, the distribution of PAHs in the Allende meteorite was analyzed. Spectra from freshly fractured interior surfaces of the meteorite showed that PAH concentrations are locally high compared to the average concentrations found by wet chemical analysis of pulverized samples. The data suggest that the PAHs are primarily associated with the fine-grained matrix, where the organic polymer occurs. In addition, highly substituted PAH skeletons were observed. Interiors of individual chondrules were devoid of PAH at the detection limit (≈ 0.05 ppm).

7.2 Laser Ablation in an Atmosphere

In this section, we consider diagnostics for ablation processes which occur in an ambient atmosphere, either terrestrial or extraterrestrial, or a synthetic atmosphere such as an inert gas environment. The pressure need not be atmospheric; we will consider here all laser-ablation processes which take place in an environment of an added background gas, even at reduced pressure. The reasons for this separation are due to the distinct physical processes which are fostered (or mitigated) when ablation occurs in a background gas. These distinctions will become abundantly clear presently. However, first it seems appropriate to address the question of why one would want to perform laser ablation in an atmosphere. There are at least two practical reasons. First, laser ablation can be used as an analytical technique (or an element of one) for field and/or remote testing of samples in terrestrial or extraterrestrial environments. In this case, laser ablation is performed in an atmosphere because there is no expedient alternative. Second, laser ablation in a background gas promotes gas-phase reactions between target atoms and the background gas as well as between the target atoms themselves. These interactions can be tailored to produce novel species and materials for basic study, for chemical analysis, and for application in advanced devices. These justifications are further considered below.

The diagnostics which we will consider here can be used to determine plume composition (in some cases quantitatively), velocity, ionization fraction, temperature, and blast-wave characteristics. This, in turn, facilitates a thorough understanding of the ablation process. This knowledge is necessary for the development of new laser ablation-based analytical methods and materials processing techniques.

7.2.1 Physical Processes Unique to Ablation in an Atmosphere

Before considering the relative merits of diagnostic techniques which have been developed to study the expansion of laser-ablated species into an atmosphere, it is instructive to consider how the physics of laser ablation into an atmosphere differs from that in vacuum, a subject treated in depth in the previous section. There are three major differences between laser ablation into vacuum and into a backing gas. First, laser ablation into an atmosphere of sufficient density produces a shock wave. The shock wave is formed by the piston-like action of the quickly expanding ablated material pushing outward on the backing gas [7.50]. As the shock expands, more background gas is swept up by the shock front, and since the laser pulse delivers a finite amount of energy, the expansion velocity decreases with increasing distance from the target [7.51]. This type of non-steady shock wave is referred to as a blast wave and is produced when a large amount of energy is deposited in a small volume with an atmosphere present to support the wave [7.52]. The wave moves symmetrically outward at very high velocity slowly decelerating to Mach 1 as the blast wave looses energy as shown in Fig. 7.3. Also depicted in Fig. 7.3 is the fact that the rate of deceleration increases for increasing pressure since the wave must push more gas at the higher pressure. *Zel'dovich* and *Raizer* [7.52] have developed a relationship for the motion of the shock front in spherical geometry which is given by (7.9)

$$R = \xi_0 (E_0/\varrho_0)^{1/5} t^{2/5} \ . \tag{7.9}$$

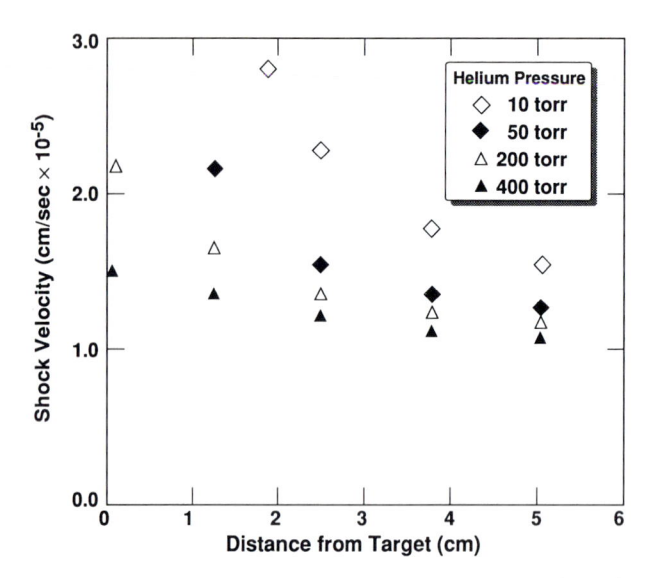

Fig. 7.3. Blast-wave velocity vs distance from an iron target for various pressures of helium backing gas. Note that the velocity decelerates to Mach 1 with increasing distance from the target. The data was obtained using the laser-deflection technique

In (7.9), R is the radius of the hemispherical blast wave front, ξ_0 is a constant which depends on the heat capacity of the background gas, E_0 is the energy deposited by the laser, ϱ_0 is the density of the unperturbed background gas, and t is the time [7.52]. *Hall* and *Bobin* et al. [7.53, 54] have confirmed the efficacy of this equation by measuring the velocity of the shock front produced by laser ablation of a metal surface in a low pressure atmosphere. From (7.9) one may obtain an equation for the velocity of the expanding wave as a function of R. This formula is given by

$$dR/dt = v = c/R^{-3/2}. \tag{7.10}$$

In (7.10), c is a collection of constants consisting of ξ_0, E_0, and ϱ_0. *Sappey* et al. [7.55] have measured the blast-wave velocity produced by excimer-laser ablation of iron targets in 10 Torr of argon. Using shadowgraphy and HeNe deflection techniques they obtain an exponent of -1.2 by fitting the data to (7.10). They have observed blast-wave velocities as high as Mach 27 using shadowgraphy in iron laser-ablation experiments with 10 Torr of argon backing gas.

The second difference between ablation into a vacuum and an atmosphere involves the extent of plasma formation in the plume. The majority of the literature on the subject of plasma formation in laser-ablated plumes deals with ablation in vacuo, although *Hermann* et al. [7.56] have recently measured electron density as a function of helium backing-gas pressure in ablation experiments with titanium targets. *Dreyfus* and *Phipps* [7.57] have written a recent review article on the subject of plasma formation in laser-ablation experiments. As a result of the paucity of information on laser ablation in a background gas, they consider ablation into a vacuum in the greatest detail; however, this information can be used to predict what effects a backing gas will have on the ionization fraction of the plume.

The salient factors determining the extent of plasma formation in vacuum are: the laser pulse length, the laser wavelength, and the intensity [7.57]. Plasma ignition is governed by the universal relationship given in (7.11):

$$I\sqrt{\tau} \geq B \ . \tag{7.11}$$

Here, I is the laser intensity in W/cm^2 and τ is the laser pulse length in seconds. B has been found to be a constant, 8×10^4 Ws$^{1/2}$/cm^2, for over a five order of magnitude range in pulse lengths (1 ms to 10 ns) and for laser wavelengths from 0.25–10 µm. The plasma ignition threshold occurs at approximately $B_p = 1/2 \ B$. The initiating electrons are most likely produced in the plasma-ignition event via multiphoton ionization of atomic and molecular species in the plume, Penning ionization of species in excited electronic states in the plume, and/or thermionic emission from the laser-heated surface. The relative importance of these three mechanisms is difficult to assess, and parameters such as wavelength influence the significance of each mechanism strongly. If the primary electrons were not present, the threshold for plasma formation would be much higher, on the order of 100 GW/cm^2 rather than 1 GW/cm^2 [7.57].

For a sufficiently long and intense ablation pulse, the latter part of the pulse couples to the primary electrons in the plasma above the surface causing heating by Inverse Bremsstrahlung (IB) [7.57]. If IB heating is appreciable, the excited electrons will dissociate any molecular species ablated from the target and electron impact ionize a large fraction of the atoms, thereby increasing the ionization fraction and promoting further absorption of the laser. This cycle produces a cascade effect which serves to increase plasma formation dramatically once threshold is achieved. The effect can be so dramatic that eventually all of the ablation laser energy may be coupled into the plasma above the surface [7.56, 57]. The wavelength of the ablation laser is important in determining the importance of this ionization mechanism because coupling to the plasma by IB is more efficient in the red part of the spectrum than in the blue, having a λ^2 dependence on wavelength. For UV-ablation wavelengths such as those produced by excimer lasers (e.g. KrF and ArF), direct photoionization or multiphoton ionization of the vapor atoms or molecules may dominate other ionization processes due to the large photon energy involved [7.57].

A backing gas alters the situation described above in the following way. The amount of laser energy which can be coupled to the surface is limited by the breakdown threshold for the experimental atmosphere. If the breakdown threshold is low, most of the energy from the laser pulse may be coupled into the buffer gas rather than the target. This effect would dramatically reduce the ablation rate. For higher breakdown thresholds, limited ionization of the backing gas may occur fostering ionization of the target atoms via charge-exchange reactions. In addition, for argon or helium backing gases, Penning ionization of the target atoms by the Ar or He metastables excited by the ablation pulse may occur [7.57].

In addition, the background gas alters the expansion dynamics of the plume. Deceleration of the ablated species [7.58], condensation of target species [7.58], redeposition, turbulence [7.59], and reaction of target species with the backing gas [7.60, 61] are all promoted or caused by the presence of an atmosphere in the ablation process. As mentioned above deceleration of the target species is the most obvious effect of background gas. The effect is quite dramatic. *Sappey* and *Gamble* [7.58] have observed the Cu atom velocity from laser-ablated copper targets to decrease from approximately 1×10^6 cm/s to 3000 cm/s in the first 100 ms after the ablation event with only 10 Torr of helium backing gas. Even more amazing is the fact that the majority of the deceleration takes place in the first microsecond [7.58].

As a result of this dramatic deceleration, a high density of the target species is maintained for rather long periods of time. High densities, in turn, promote condensation of the target species by the following mechanism. The formation of particulate ultimately relies on the formation of the dimer of the target species, assuming a completely atomized plume. The dimer is produced by a collision between two target atoms which occurs with a rate proportional to the square of the target atom density. Therefore, the backing gas promotes condensation of the target species by maintaining the high target-atom densities necessary for dimerizing collisions to occur. However, this is not the only function of the backing gas. To form a stable dimer or *small* cluster, a third body is required to remove the relative translational energy of the collision pair and to remove some of the energy given up by the formation of the new bond. This three-body formation mechanism is vital to the production of small clusters which do not have sufficient internal degrees of freedom to randomize the energy given up by the formation of a new bond when an atom adds to the cluster. In general, the backing gas is present in great excess of the target-atom density in laser ablation–condensation experiments; therefore, the backing gas nominally provides the third body. Finally, the backing gas cools the "protocluster" which is formed in highly excited rovibration states. Depending on the internal energy content and size of the protocluster, significant cooling may have to occur before the cluster can add an additional atom(s). Thus condensation depends in a highly nonlinear way on the pressure of the backing gas [7.58].

Vortex formation and turbulence are relatively familiar fluid-dynamic phenomena which are caused by viscous effects in a flow field. A familiar example of the result of vortex formation is the characteristic shape of mushroom clouds which are formed by ground-based explosions in an atmosphere. Laser ablation is, in many respects, similar to a ground based explosion, so it is not surprising that *Gilgenbach* and *Ventzek* [7.59] have observed what they believe to be turbulent structure in aluminum-ablation experiments with background atmospheres consisting of 760 Torr of argon or air. *Sappey* and *Gamble* [7.58] have observed the formation of mushroom clouds in Planar Laser-Induced Florescence (PLIF) images of Cu and Cu_2 when ablating copper into a background gas consisting of 25–100 Torr of helium. One result of vortex formation and/or turbulence is to facilitate mixing between target species and the backing gas which may promote condensation of the target species and/or reaction of the target species with the backing gas.

Gas-phase reactions between constituents ablated from $YBa_2Cu_3O_{7-\delta}$ HTS and O_2 have been shown by several research groups to be important for the production of high-quality thin films of HTSs. *Otis* and *Dreyfus* [7.69] have studied these reactions in detail and find that the production of CuO is maximized with a backing-gas pressure of 100 mTorr of O_2. This suggests that CuO is critical for production of films with the proper stoichiometry. On the other hand, *Dye* et al. [7.60] have found that the reaction between ablated Y atoms and O_2 reaches a limiting value with 400 mTorr of O_2.

Similarly, *Gilgenbach* and *Ventzek* [7.59] have observed significant differences in laser-ablated plumes of Al depending on whether the backing gas is atmospheric-pressure argon or air. They attribute these differences to oxidation reactions between Al and O_2 which are highly exothermic. In support of this conclusion, independent experiments which monitor dispersed AlO chemiluminescence have confirmed that AlO is produced as a reaction product when an Al target is ablated into an oxygen containing atmosphere [7.62]. Several research groups have utilized the *Smalley* cluster source or variations of it with reagents added to the stagnation gas to produce exotic species for spectroscopic characterization. The Smalley apparatus consists of a laser-ablation source coupled with a supersonic pulsed valve and is used to form clusters of refractory species [7.63]. For example, *Richtsmeier* et al. [7.64] have modified the original *Smalley* design and developed a cw flow reactor to study the reactivity of metal clusters with various reagents.

7.2.2 Diagnostics for Laser Ablation in an Atmosphere

This portion of the chapter is divided into three sections each of which deals with a particular aspect of ablation in an atmosphere: blast-wave diagnostics, plasma-formation diagnostics, and diagnostics for condensation or reactions taking place in the plume. The section on blast wave diagnostics focuses on methods for imaging the shock front and measuring its velocity. The second section dealing with plasma formation considers methods for determining the onset of the plasma and measuring plasma parameters such as electron density and temperature as well as ionization fraction. The final section treats diagnostics used in the determination of reaction and condensation mechanisms. In this section we include methods which can be used for the determination of density, temperature, and velocity of atomic and molecular constituents in the plume. In some cases the same diagnostic technique can be used to measure several quantities. In these situations, the details are described only once; however, the applications are discussed in more detail. In some cases, diagnostic techniques which are well developed for use in other fields will be suggested for use in laser-ablation experiments due to the lack of adequate diagnostics for the quantity desired. The list of diagnostics which we present here is not exhaustive; however, most of the better known laser-ablation diagnostics are discussed.

a) Blast-Wave Diagnostics

The production of blast waves during laser ablation may be monitored by any of the following diagnostic techniques: interferometry, Schlieren photography, shadowgraphy, and laser-beam deflection. These techniques share a common feature in that each is sensitive to changes in refractive index; however, each has attractive features peculiar to it which makes it particularly useful for some types of measurements and less desirable for others. Interferometry, Schlieren photography, and shadowgraphy are sensitive to the refractive index, the first derivative of the refractive index, and the second derivative of the refractive index, respectively, and all three yield two-dimensional fields. However, the Schlieren and shadowgraphy techniques are inherently difficult to quantify. Interferometric measurements are relatively easy to quantify but can be difficult to implement. Conversely, beam-deflection experiments are easily quantifiable and easy to implement but yield information at only one point in the sample and along a line of sight similar to absorption experiments.

Schlieren photography is a well known density- or refractive-index sensitive technique for imaging high-speed flows. The technique has been applied in aerospace engineering for decades to aid in the design of aircraft. A Schlieren experiment gives rise to signals which are the first derivative of the refractive index dn/dx. The Schlieren method is sensitive to refractive-index gradients normal to a knife edge placed at the focus of light exiting the test section. Since the blast wave in laser-ablation experiments creates regions of high and low density, the Schlieren technique is very useful for imaging it. Although a laser light source is not necessary for this experiment, it is convenient in that the light is intense and easily manipulated to the desired size. *Ventzek* et al. [7.50] have used the Schlieren technique to image the blast wave created by ablation of polymers in various gases at atmospheric pressure. Also visible in their Schlieren images are regions of high density behind the shock wave ascribed to material ablated from the surface. *Sappey* et al. [7.55] have used Schlieren photography to image the blast wave created in ablation experiments of iron targets with various pressures of argon backing gas.

Shadowgraphy is quite similar to the Schlieren technique in experimental setup, although shadowgraphy is somewhat less complex and easier to implement. There is no knife edge used in shadowgraphy. *Srinivasan* et al. [7.31] have used shadowgraphy (although they do not refer to it as such) to image the blast-wave front created by the ablation of polymer surfaces in air at atmospheric pressure. They observe a sharp shock front as well as less intense "shocklets" which they do not mention. In addition, their data show a less well defined region behind the front which is apparently material ablated from the surface

[7.31]. *Sappey* et al. [7.55] have used shadowgraphy to image the blast wave created by ablation of an iron target in various pressures of argon backing gas. They observe a main shock front as well as less intense "shocklets" as in the data of *Srinivasan* et al. [7.31]. The origin of the "shocklets" is not clear, although they seem to be a fairly general phenomenon in laser-ablation experiments performed in a backing gas.

Interferometry is another technique which is potentially useful for imaging the blast wave created in laser-ablation experiments, although to our knowledge, it has never been used for this purpose. However, *Horton* and *Gilgenbach* [7.65] have used holographic interferometry to image the shock wave and reduced-density channel created by CO_2-laser-induced breakdown of atmospheric pressure He gas. Interferometry has the definite advantage that it gives the density field directly as opposed to Schlieren techniques and shadowgraphy which give, respectively, the first and second derivatives of the density field.

Laser beam deflection experiments measure the displacement of a laser beam on a position sensitive detector caused by refractive-index gradients in the sample under study. The gradients in refractive index may be caused by temperature, plasma formation, or simply density changes in the sample. Since the blast wave created in ablation experiments causes a density gradient, the laser beam deflection technique is sensitive to it. *Sappey* et al. [7.55] have used HeNe-beam deflection to measure the blast wave velocity in iron laser ablation experiments as have *Petzoldt* et al. [7.66] in laser-damage experiments in air and *Ventzek* et al. and *Sell* et al. [7.50, 67] in laser ablation experiments of polymers. Figure 7.3 shows a graph of shock velocity versus distance from the target for four pressures of helium backing gas, 10, 50, 200, and 400 Torr. Notice that the velocities asymptotically approach the speed of sound in helium at that pressure. Note also that the shock speed decelerates more quickly at higher pressures as predicted by (7.9, 10). *Petzoldt* and *coworkers* [7.66] develop a relationship between the amplitude of the signal and the energy contained in the acoustic pulse. This relationship is used to correlate the energy contained in the blast wave with the damage done to the surface.

b) Optical Diagnostics for Monitoring Plasma Formation

Optical diagnostics for studying the onset of plasma formation in laser ablation experiments have been developed to measure quantities such as electron (ion) density, electron temperature, and ion translational temperature. Techniques such as laser beam deflection [7.68], Thomson scattering [7.69], ion spectroscopy (absorption, emission, or laser induced fluorescence) [7.56, 70], as well as Stark broadening of H atom emission lines [7.56] have been used to measure such quantities. This section presents a review of techniques which have been used to measure parameters associated with laser-ablated plasmas.

A very useful technique for monitoring plasma formation in laser-ablation experiments is the laser-beam deflection technique which has also been used to measure blast-wave velocities as described before. *Enloe* et al. [7.68] describe in detail the application of HeNe-deflection measurements to the study of plasma formation in laser-ablation experiments. They note that for plasmas the change in refractive index δn is always negative and proportional to the electron density n_e, and is given by

$$\delta n = -(e^2/2\pi m f^2)n_e = -K_p n_e , \tag{7.12}$$

where e is the electron charge, m is its mass, and f is the laser frequency. Therefore, $K = 1.79 \times 10^{-22}$ cm^3 at the HeNe-laser frequency. For neutral particles, δn is positive and is given by:

$$\delta n = (n_0 - 1/n_0)_{STP} n_n = K_n n_n . \tag{7.13}$$

Figure 7.4 shows a HeNe-deflection trace obtained by *Enloe* et al. [7.68] while ablating graphite in vacuum although the technique will also work for elevated pressures. According to *Enloe* et al., the positive and negative deflections correspond to changes in refractive index caused by electrons and neutrals, respectively, according to (7.12, 13). Data such as that in Fig. 7.4 can be used to measure ionization onset, electron (ion) density, neutral density as well as neutral and ion velocities.

Since the onset of the plasma is characterized by changes in refractive index due to the production of electrons (7.12), all of the refractive index sensitive techniques discussed before (Schlieren, shadowgraphy, interferometry, and laser deflection) should be sensitive to plasma formation. For example, *Walkup* et al. [7.71] have used a Michelson interferometer to monitor plasma formation in laser ablation experiments of several insulators, semiconductors, and metals. The Michelson interferometer measures changes in refractive index caused by the plume passing through the laser beam in one arm of the interferometer. A free-electron gas has a refractive index less than 1.0 while neutrals have an index of refraction greater than 1.0 (7.12, 13). Thus, the intensity of light transmitted by

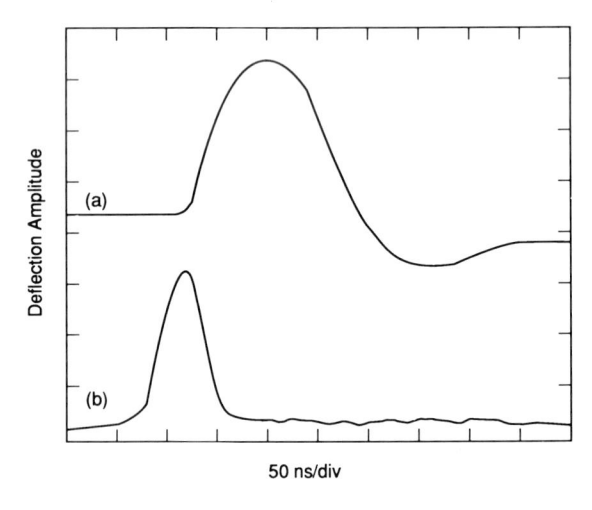

Fig. 7.4. (a) Laser-deflection trace obtained by *Enloe* et al. [7.68] for laser ablation of a graphite target in vacuum. Trace (b) is the temporal profile of the ablation-laser pulse. The positive deflection in trace (a) is due to electrons; the negative deflection is due to neutrals. Data such as depicted in Fig. 7.4 can be used to measure electron and neutral density and velocity

the interferometer shows opposite-sign deflections depending on whether the medium causing the change in refractive index is neutral or ionized. *Walkup* et al. [7.71] estimate a very high ionization fraction of approximately 10% for a wide range of inorganic materials such as Al, Cu, Si, Ge, and MgO. They observe very different behavior for ablation of organic polymers typically observing an *upper limit* to the ionization fraction in such plumes of 0.1–1%. Finally, *Walkup* and coworkers observe a strong correlation between the electron–ion signal from the interferometer and optical emission from neutral excited states. They conjecture that the excited states are formed by electron impact excitation and ion electron recombination which would explain the correlation between plasma formation and optical emission [7.71].

Another technique sensitive to plasma parameters is Thomson scattering which yields data on electron density and ion velocity. The physics of Thomson scattering involves the acceleration of the plasma electrons by the electric field of the laser beam. The accelerated charges emit light at approximately the frequency of the laser. The details of the lineshape of the scattered light contain information about electron temperature [7.69]. Specifically, the scattered radiation exhibits a bimodal lineshape the splitting of which is $2k_{ia}c_s$. The wave vector of the ion-acoustic wave k_{ia}, is given by (7.14), where θ_{obs} is the angle between the wave vectors of the probe and scattered radiation:

$$k_{ia} = (4\pi/\lambda_{probe}) \sin(\theta_{obs}/2) \ . \tag{7.14}$$

The sound speed c_s is then calculated from the measured splitting and the electron temperature is inferred. The center line of the bimodal peaks exhibits a Doppler shift, $k_{ia} U_{drift}$, from the probe frequency and thus yields a directionally dependent drift velocity [7.69]. The total energy scattered is proportional to the electron density and can be measured with a calibrated light collection system. The cross section for Thomson scattering is given by:

$$\sigma_T = (e^2/mc^2)\mathbf{n} \times (\mathbf{n} \times \varepsilon_0) \ , \tag{7.15}$$

where $\mathbf{n}$ is a unit vector directed from the charge to the point of observation and $\varepsilon_0 = E_0/|E_0|$, and all other parameters assume their normal significance. As seen in (7.15), the scattering cross section is inversely proportional to the mass of the scatterer. Therefore, the contribution to the signal from ions is vanishingly small compared with the contribution from electrons. The value of this cross section is very small ($\sigma_T = 7.952 \times 10^{-26}$ cm^2) and is a major limitation on the usefulness of Thomson scattering in application. However, *Tracy* and *coworkers* [7.69] have succeeded in using Thomson scattering to obtain electron density, drift velocity, and temperature in laser-ablation experiments of aluminum targets in vacuum. The technique should also work for slightly elevated pressures, although, presumably, Rayleigh scattering from the background gas will eventually dominate Thomson scattering.

Another optical technique suitable for measuring plasma parameters is Stark broadening of hydrogen Balmer-series emission lines. *Cremers* et al. [7.72] have used the breadth of the H_α line at 656.28 nm as a diagnostic for electron density in cw CO_2-laser-sustained Xe plasmas (seeded with a small amount of H_2). The breadth of the H_β line at 486.13 nm gives a more accurate determination of electron density but could not be used in that study because of interference from Xe (I) lines as well as the six-fold decrease in the emission intensity on the H_β line compared to the H_α line. *Hermann* et al. [7.56] have used Stark broadening of hydrogen-atom lines to measure electron densities as high as 10^{18} cm^{-3} in Ti-ablation experiments in helium backing gas.

Finally, conventional spectroscopy (absorption, emission, and LIF) on ions in laser-ablated plumes has been used to measure ion velocity and density. These spectroscopic techniques are also useful for determining the onset of ionization and plasma formation in the plume. For example, *Dreyfus* [7.70] has used laser-induced fluorescence to detect the $^3D_3$ metastable state of the Cu^+ ion to determine the onset of plasma formation in laser-ablation experiments. *Hermann* et al. [7.56] use emission spectroscopy to determine the time dependence for Ti^+, Ti^{2+}, and Ti^{3+} as well as He^+ in laser-ablated Ti plumes in helium backing gas [7.55]. *Sappey* et al. have used emission spectroscopy to determine the excitation temperature of Fe atoms and to estimate the ionization temperature in laser-ablation experiments occurring in various pressures of backing gas assuming local thermodynamic equilibrium. *Walkup* et al. [7.71] have observed a close correspondence between optical emission from neutral excited states and plasma formation for many inorganic substances suggesting that optical emission may be used as a diagnostic for plasma formation in many cases.

c) Density, Temperature, and Velocity Diagnostics

In this section, techniques for measuring the density, temperature, and velocity of atoms and molecules resulting from laser ablation into a backing gas are discussed. These quantities are of paramount importance in understanding the nature of the ablation event. The density and temperature are especially important in characterizing the reactive processes which may occur in such plumes. Unfortunately, these quantities are also the most difficult to measure. We will begin by considering density sensitive diagnostic techniques.

Density Diagnostics. Perhaps the most straightforward density diagnostic is the classical absorption experiment. *Geohegan* and *Mashburn* [7.73] have used transient optical-absorption spectroscopy to study the neutral and ion transport in ablation of HTS material. Their experimental apparatus consists of a pulsed high-intensity Xe lamp and a 1 m monochromator with a PMT detector. The light from the lamp is passed through the sample to the monochromator where it is dispersed and detected by the PMT. They monitored the temporal profiles of Y, Ba, Cu, and Ba^+ at various distances from the target. Although this technique could be used to measure the density of these plume species, *Geohagen* and *Mashburn* limited their study to the determination of the velocity of these plume species. One drawback of the absorption technique is that it integrates over the light-path length through the sample. Therefore, to calculate absolute number densities, one must know the distribution of probed species over the path length. This can be accomplished by the use of PLIF, as described below.

Sappey and *Gamble* [7.74] have used laser-beam absorption to measure the density of Cu atoms in laser-ablation experiments of copper targets in 10 Torr or less of helium backing gas. They develop an expression for the Cu-atom absorption coefficient at line center appropriate for their experimental conditions and determine the Cu density to be approximately 10^{14} cm^{-3} with 10 Torr of helium backing gas, assuming a top-hat distribution of Cu atoms in the plume. In a subsequent paper, *Sappey* and *Gamble* [7.58] determine the actual distribution of Cu atoms in the plume using PLIF. The actual distribution of Cu atoms is decidedly peaked in the center of the plume which causes the 10^{14} cm^{-3} value to be slightly low. They also observe that at backing gas pressures higher than 10 Torr, pressure and opacity broadening cause the spectral lineshape to become Lorentzian and assumptions made to calculate the absorption coefficient at line center for pressures less than 10 Torr are no longer valid [7.58]. To circumvent problems associated with spectral-line broadening in the determination of plume density, *Sappey* and *Gamble* [7.75] have recently demonstrated the use of a somewhat obscure technique for the quantitative measurement of atomic-plume constituents at elevated pressures. This technique, Hook spectroscopy, is an interferometric/spectroscopic method which has been used primarily for the measurement of atomic oscillator strengths [7.76]. The beauty of the technique is that it is independent of line-broadening mechanisms. One needs to know only the oscillator strength of the transition to obtain a density. Therefore, Hook spectroscopy is ideally suited for density measurements in laser-ablation experiments with high backing-gas pressures and/or large optical depth.

Enloe et al. [7.68] use laser-beam deflection as a quantitative diagnostic for both ions and neutrals, as mentioned in the plasma diagnostics section. Their technique relies on refractive-index gradients caused by the neutral species in the plume altering the direction of a cw probe-laser beam (usually a HeNe). The magnitude of this deflection is measured by a position-sensitive detector and can be related directly to the concentration of neutral species in the plume.

In principle, LIF or PLIF can be used to determine the density in laser-ablated plumes. In practice, this is an extremely difficult task, especially at elevated pressures and high densities. The light-collection system must be calibrated which can be difficult; however, other information such as quantum yield is necessary for accurately determining the density. The quantum yield is, in turn, influenced by quenching, the degree to which the laser saturates the transition, amplified spontaneous emission from the excited state, and line-broadening mechanisms. These problems cause the LIF and PLIF techniques to be of limited use for density determination in laser-ablated plumes at elevated background pressures.

Temperature Determination. Temperature measurement in laser-ablated plumes has generally been determined by rotationally resolved LIF excitation scans of electronic transitions in small diatomic molecules [7.58, 77], by measuring vibrational-state populations [7.78], or by measuring the Doppler width of atomic lines in the plume [7.78]. The latter technique has been used by *Dreyfus* to measure the temperature of ablated-copper plumes with no backing gas. *Dreyfus* observes temperatures near 3000 K which is believed to be the maximum surface temperature attained during the ablation pulse. In addition, *Dreyfus* [7.78] has measured the vibrational and rotational temperature of C_2 ablated from graphite targets with no added backing gas. They observe a Boltzmann distribution for both vibrational and rotational distributions characterized by the same temperature, 3600 K, the estimated sublimation temperature.

Sappey and *Gamble* [7.74] have used rotationally partially resolved excitation scans of the $Cu_2 A \leftarrow X$ (0,0) transition to measure the temperature of condensation produced Cu_2 in copper plumes with various pressures of added helium backing gas from 10–100 Torr. As shown in Fig. 7.5, a temperature is obtained which is very nearly 300 K, the temperature of the helium backing gas for all conditions studied. In a subsequent paper, *Sappey* and *Gamble* [7.58] reason that the uniform 300 K temperature is due to the physics of the condensation process which occurs to produce vibrationally excited Cu_2. Many

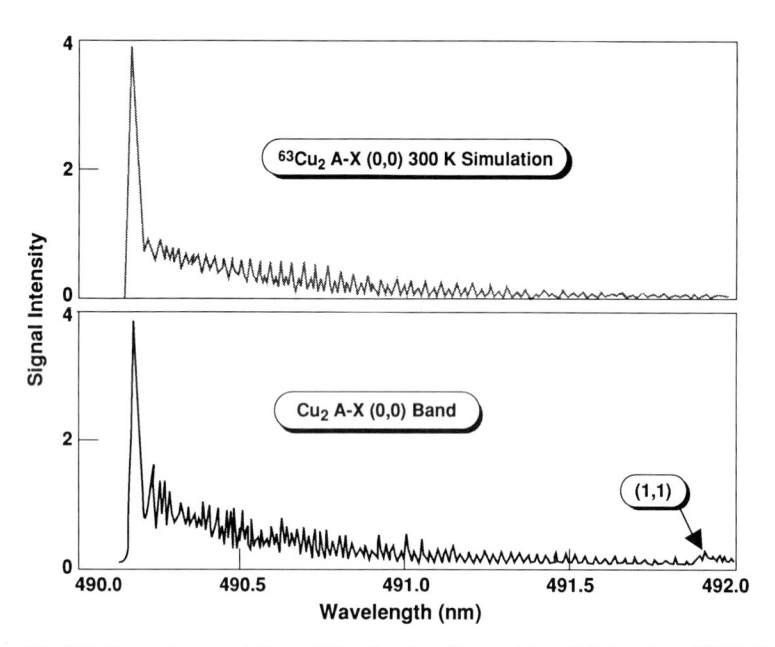

Fig. 7.5. Computer simulation of the $Cu_2 A \leftarrow X$ transition (0,0) band at 300 K. Only the 63-63 isotopomer is included in the simulation (*top*). Experimental LIF spectrum of condensation-produced Cu_2 (*bottom*). The congestion present in the experimental spectrum which is not in the simulation is due to the 63-65 and 65-65 isotopes of Cu_2. The (1, 1) band is seen in the experimental spectrum but not included in the simulation. The Cu_2 temperature is approximately 300 K

collisions are necessary to vibrationally deactivate the Cu_2 to $v = 0$ of the ground state which is the level probed in the LIF experiment. These same collisions rotationally relax the distribution to the ambient temperature of the He backing gas, 300 K [7.58]. *Sappey* et al. judge the vibrational temperature of the condensation-produced Cu_2 to be nearly 300 K from the relative intensities of the $A \leftarrow X$ (0,0) and (1,1) bandheads after taking into account the pre-dissociation of $v' = 1$ of the A state of Cu_2.

The electronic excitation temperature is easily measured in laser-ablated plumes; however, one should not expect the electronic degree of freedom to be in equilibrium with the gas kinetic temperature. *Hermann* et al. and *Sappey* et al. have used spatially and temporally resolved emission and Boltzmann analysis to determine electronic excitation temperatures for laser ablation of titanium and iron surfaces in helium backing gas, respectively [7.56]. *Tremblay* et al. [7.79] have used temporally resolved emission from ablated-iron plumes to determine electronic excitation temperatures.

d) Ablated Material Velocity Determination

Velocity measurement in laser-ablation experiments has been accomplished by LIF [7.74, 78, 80], PLIF [7.58, 81], laser-beam deflection [7.68], transient absorption [7.73], spectrally, spatially and temporally resolved emission [7.56], fast framing photography [7.82], and dye-laser resonant absorption [7.59].

LIF is a convenient means of measuring the velocity of species in a laser-ablated plume. LIF is species specific and is normally implemented as a point diagnostic. TOF measurements using the LIF signal are typically used to measure velocity. At elevated pressures, the LIF signal must be monitored at several different target–probe-beam separations because the backing gas causes the ablated material to decelerate rapidly. Such TOF measurements may also be implemented using direct absorption spectroscopy [7.73]. *Qu* et al. [7.80] have used a cw ring dye laser to excite LIF in laser-ablated uranium plumes and produce uranium TOF curves. The advantage of this technique is that an entire time-of-flight profile can be obtained for a single ablation event by digitizing the PMT signal as the species of interest passes through the cw laser. The use of pulsed-excitation lasers for time-of-flight measurements allows only a single point on the time-of-flight curve to be obtained for a single ablation event. Thus the production of a complete time-of-flight curve is quite time intensive and also depends on good shot-to-shot reproducibility.

PLIF is a variation of the normal LIF experiment; however, in a PLIF experiment, the laser is fashioned into a thin sheet which is passed through the sample to excite an atomic or molecular species of interest. Fluorescence is captured by a two-dimensional detector such as a CCD camera to yield a two-dimensional density field of the species excited within the sample. In addition to being useful for simple visualization, PLIF has been used to measure the velocity of laser-ablated plumes in vacuum and in a backing gas as well as the diffusion velocity of species ablated into a backing gas [7.81]. Dye-Laser

Resonant Absorption (DLRA) is the two-dimensional analog of direct absorption. *Gilgenbach* and *Ventzek* [7.59] have used DLRA to image ablated aluminum plumes in argon and oxygen containing atmospheres to obtain velocity. However, the DLRA experiment integrates over the path length of the laser through the sample, as in normal absorption experiments, making this technique somewhat less useful than PLIF for some applications.

Another velocity measurement technique utilizing LIF or absorption involves the determination of the Doppler shift of a transition. In such an experiment, the wavelength of a spectral peak in the expanding ablation plume is measured relative to the wavelength of the same transition in a sample exhibiting only thermal motion. The velocity is given by:

$$v = v_0[1 + (v/c)\sin\theta] \; . \tag{7.18}$$

Here, v_0 is the center frequency of the excited transition, v is the velocity of the sample, θ is the angle between the laser-propagation direction and the sample-velocity vector, c is the speed of light, and v is the Doppler-shifted frequency. Doppler shift is commonly used for velocity determination in hypersonic flows. For example, *Liebeskind* et al. [7.83] have performed LIF experiments exciting the hydrogen Balmer α transition to measure hydrogen-atom velocity in an hydrogen arcjet.

Laser-beam deflection has been used by *Enloe* et al. [7.68] to determine ablated material velocity using carbon as the target material. They develop a simple relationship between the velocity of the ablation-produced ions or neutrals and the timing of the maximum deflection with respect to the ablation event. Laser-beam deflection has many attractive features such as ease of implementation; however, one important disadvantage is the lack of species selectivity. Several researchers have used spectrally, spatially, and temporally resolved emission to measure the velocity of excited-state atomic species in laser-ablation experiments [7.56]. This technique is both easy to implement and species selective; however, it is limited by its reliance on light from plume emission which typically lasts for $\leq 10\,\mu s$ after the ablation event. *Dyer* and *Sidhu* [7.51] have demonstrated the utility of Fast-Framing (FFP) and streak-camera Photography for measuring the velocity of electronically excited species from ablation of polymer materials. In FFP, a short exposure series of pictures is obtained documenting the expansion of a single-ablation plume at well-defined intervals. The self-luminosity of the plume provides the light source and the exposure time is short enough to freeze the motion of the plume. If the length scale is accurately calibrated, a velocity can easily be calculated from the fast-framing-camera record. Streak photography is similar to framing photography except that the camera intensifier remains on for the duration of the luminous plume. A streak-like record is produced which gives a continuous position versus time history of the luminous component of the plume.

The preceding sections described diagnostic techniques suitable for probing laser-ablated plumes occurring in a background atmosphere. Techniques are described which measure blast-wave parameters, ionization fraction, velocity, density (composition), and temperature. These diagnostics yield fundamental information on plume physics and chemistry which have been used in some cases to develop analytical and materials processing techniques. While much has been learned in the laser–materials interaction field, the application of new diagnostic techniques is sure to lead to new discoveries of both a fundamental and applied nature.

References

7.1   J.F. Ready: *Effects of High-Power Laser Radiation* (Academic, New York 1971) p. 433
7.2   L.J. Radziemski, D.A. Cremers: In *Laser-Induced Plasma and Applications* (Dekker, New York 1989)
7.3   H. Sankur, J.T. Cheung: Appl. Phys. A **47**, 271 (1988)
7.4   W. Zapka, W. Ziemlich, A.C. Tam: Appl. Phys. Lett. **58**, 2217 (1991)
7.5   R.C. Dye, S.R. Foltyn, N.S. Nogar, X.D. Wu, E.J. Peterson, R.E. Muenchausen: In *Proc. Laser Processes for Microelectronic Applications* (Electrochem. Soc., New York 1992)
7.6   L. Moenke-Blankenburg: *Laser Microanalysis* (Wiley, New York 1989) p. 288.
7.7   D.J. Burgess, R. Viswanathan, I. Hussla, P.C. Stair, E. Weitz: J. Chem. Phys. **79**, 5200 (1983)
7.8   R.C. Estler, E.C. Apel, N.S. Nogar: J. Opt. Soc. Am. B **4**, 281 (1987)
7.9   D.A. Cremers: Appl. Spectrosc. **41**, 572 (1987)
7.10 H.S. Kwong, R.M. Measures: Anal. Chem. **51**, 428 (1979)
7.11 E.R. Denoyer, K.J. Fredeen, J.W. Hager: Anal. Chem. **63**, 445A (1991)
7.12 R.J. Cotter: Anal. Chim. Acta **195**, 45 (1987)
7.13 R.S. Houk, S.C.K. Shum, D.R. Wiederin: Anal. Chim. Acta **250**, 61 (1991)
7.14 R.J. Conzemius, J.M. Capellen: Int. J. Mass Spectrom. Ion Phys. **34**, 197 (1980)
7.15 C.H. Becker: *Methods Mech. Prod. Ions Large Mol.*, Nato ASI Ser. B. Vol. **269**, 293 (1991)
7.16 K. Dittrich, R. Wennrich: Progr. Anal. At. Spectrosc. **7**, 139 (1984)
7.17 K.W.D. Ledingham: Anal. Proc. (London) **28**, 413 (1991)
7.18 G.R. Van, C. Xhoffer: J. Anal. At. Spectrom. **7**, 81-8 (1992)
7.19 E. Denoyer, G.R. Van, F. Adams, D.F.S. Natusch: Anal. Chem. **54**, 26A (1982)
7.20 P. Rechmann, J.L. Tourmann, R. Kaufmann: Proc. of Applications of Lasers in Orthop., Dent., Vet. Med., 1991 (SPIE, 1991)
7.21 N.S. Nogar, R.C. Estler, C.M. Miller: Anal. Chem. **57**, 2441 (1985)
7.22 B. Spengler, R. Kaufmann: Analysis **20**, 91 (1992)
7.23 M. Karas, U. Bahr, U. Giessmann: Mass Spectrom. Rev. **10**, 335 (1991)
7.24 N.S. Nogar, R.C. Estler, B.L. Fearey, C.M. Miller, S.W. Downey: Nucl. Instrum. Methods Phys. Res. B **44**, 459 (1990)
7.25 R.L. Hanson, D. Brookins, N.E. Vanderborgh: Anal. Chem. **48**, 2210 (1976)
7.26 D.A. McCrery, E.B.J. Ledford, M.L. Gross: Anal. Chem. **54**, 1435 (1982)
7.27 M.R. Chevrier, R.J. Cotter: Mass Spectrm. **5**, 611 (1991)
7.28 R. Viswanathan, I. Hussla: J. Opt. Soc. Am. B3, 796 (1986)
7.29 D.J. Burgess, P.C. Stair, E. Weitz: J. Vac. Sci. Technol. A**4**, 1362 (1986)
7.30 R.C. Estler, N.S. Nogar: Appl. Phys. Lett. **49**, 1175 (1986)
7.31 R. Srinivasan, K.G. Casey, B. Braren, M. Yeh: J. Appl. Phys. **67**, 1604 (1990)
7.32 J.D. Fassett, L.J. Moore, J.C. Travis, F.E. Lytle: Int. J. Mass Spectrom. Ion Phys. **54**, 201 (1983)
7.33 G.S. Hurst, M.G. Payne, S.D. Kramer, J.P. Young: Rev. Mod. Phys. **51**, 767 (1979)
7.34 C. M. Miller, N.S. Nogar, A.J. Gancarz, W.R. Shields: Anal. Chem. **54**, 2377 (1982)

7.35 B. Schueler, R.W. Odom: J. Appl. Phys. **61**, 4652 (1987)
7.36 C.H. Becker, K.T. Gillen: Appl. Phys. Lett. **45**, 1063 (1984)
7.37 C.M. Miller, J.B. Cross, N.S. Nogar: Opt. Commun. **40**, 271 (1982)
7.38 D.P. Land, H.C.L. Pettiette, J.C. Hemminger, R.T.J. Mclver: Acc. Chem. Res. **24**, 42 (1991)
7.39 P.H. Hemberger, N.S. Nogar, J.D. Williams, R.G. Cooks, J.E.P. Syka: Chem. Phys. Lett. **191**, 405 (1992)
7.40 N.S. Nogar, R.C. Estler: Laser desorption laser ablation with detection by resonance ionization mass spectrometry, in *Lasers and Mass Spectrometry*, ed. by D.M. Lubman (Oxford, Univ. Press, New York 1990) pp. 65
7.41 R. Kelly, R.W. Dreyfus: Nucl. Instrum. Methods Phys. Res. **B32**, 341 (1988)
7.42 R. Kelly, R.W. Dreyfus: Surf. Sci. **198**, 263 (1988)
7.43 R. Kelly, J. Chem. Phys. **92**, 5047 (1990)
7.44 R.C. Estler, N.S. Nogar: J. Appl. Phys. **69**, 1654 (1991)
7.45 K.B. Jacobson, H.F. Arlinghaus, H.W. Schmitt, R.A. Sachleben, G.M. Brown, N. Thonnard, F.V. Sloop, R.S. Foote, F.W. Larimer, Genomics **9**, 51 (1991)
7.46 K.B. Jacobson, H.F. Arlinghaus: Anal. Chem. **64**, 315A (1992)
7.47 J.H. Hahn, R. Zenobi, J.L. Bada, R.N. Zare: Science **239**, 1523 (1988)
7.48 R. Zenobi, J.M. Philippoz, P.R. Buseck, R.N. Zare: Science **246**, 1026 (1989)
7.49 L.J. Kovalenko, C.R. Maechling, S.J. Clemett, J.M. Philippoz, R.N. Zare, C.M.O. Alexander: Anal. Chem. **64**, 682 (1992)
7.50 P.L.G. Ventzek, R.M. Gilgenbach, J.A. Sell, D.M. Heffelfinger: J. Appl. Phys. **68**, 965 (1990)
7.51 P.E. Dyer, J. Sidhu: J. Appl. Phys. **64**, 4657 (1988)
7.52 Y.P. Zeldovich, Y.P. Raizer: *Physics of Shock Waves and High Temperature Hydrodynamic Phenomena.* (Academic, New York 1966) pp. 93
7.53 R.B. Hall: J. Appl. Phys. **40**, 1941 (1969)
7.54 J.L. Bobin, Y.A. Durand, P.P. Langer, G. Tonon: J. Appl. Phys. **39**, 4184 (1968)
7.55 A.D. Sappey, T.K. Gamble, P.J. Wantuck, H.H. Watanabe, R. Benjamin: *Condensation Diagnostic Studies of Laser Ablated Iron Plasmas*, # LAUR-90-3258, (Los Alamos National Laboratory, 1990)
7.56 J. Hermann, L.C. Boulmer, B. Dubreuil: Appl. Surf. Sci. **46**, 315 (1990)
7.57 R.W. Dreyfuss, C.R. Phipps: Laser ablation and plasma formation, in *Laser Ionization Mass Analysis*, ed. by A. Vertes, R. Gijbels, F. Adams (Wiley, New York 1992) preprint
7.58 A.D. Sappey, T.K. Gamble: J. Appl. Phys. (1992) in press
7.59 R.M. Gilgenbach, P.L.G. Ventzek: preprint
7.60 R.C. Dye, R.E. Muenchausen, N.S. Nogar: Chem. Phys. Lett. **181**, 531 (1991)
7.61 C.E. Otis, R.W. Dreyfus: Phys. Rev. Lett. **67**, 2102 (1991)
7.62 A.D. Sappey, T.K. Gamble: Unpublished data (1991)
7.63 D.E. Powers, S.G. Hansen, M.E. Geusic, A.C. Puiu, J.B. Hopkins, T.G. Dietz, M.A. Duncan, S.P.R.R. Langridge, R.E. Smalley: J. Phys. Chem. **86**, 2556 (1982)
7.64 S.C. Richtsmeier, E.K. Parks, K. Liu, L.G. Pobo, S.J. Riley: J. Chem. Phys. **82**, 3659 (1985)
7.65 L.D. Horton, R.M. Gilgenbach: Appl. Phys. Lett. **43**, 1010 (1983)
7.66 S. Petzoldt, A.P. Elg, M. Reichling, J. Reif, E. Matthias: Appl. Phys. Lett. **53**, 2005 (1988)
7.67 J.A. Sell, D.M. Heffelfinger, P. Ventzek, R.M. Gilgenbach: Appl. Phys. Lett. **55**, 2435 (1989)
7.68 C.L. Enloe, R.M. Gilgenbach, J.S. Meachum: Rev. Sci. Instrum. **58**, 1597 (1987)
7.69 M.D. Tracy, G.J.S. De, K.G. Estabrook, S.M. Cameron: Phys. Fluids B **4**, 1576 (1992)
7.70 R.W. Dreyfus: J. Appl. Phys. **69**, 1721 (1991)
7.71 R.E. Walkup, J.M. Jasinski, R.W. Dreyfus: Appl. Phys. Lett. **48**, 1690 (1986)
7.72 D.A. Cremers, F.L. Archuleta, R.J. Martinez: Spectrochim. Acta **40B**, 665 (1985)
7.73 D.B. Geohegan, D.N. Mashburn: Appl. Phys. Lett. **55**, 2345 (1989)
7.74 A.D. Sappey, T.K. Gamble: Appl. Phys. B **53**, 353 (1991)
7.75 A.D. Sappey, T.K. Gamble: Appl. Phys. Lett. (1992), to be published
7.76 W.C. Marlow: Appl. Opt. **6**, 1715 (1967)
7.77 R.W. Dreyfus, R. Kelly, R.E. Walkup: Nucl. Instrum. Methods Phys. Res. **B23**, 557 (1987)
7.78 R.W. Dreyfus: High Temp. Sci. **27**, 503 (1990)

7.79 M.E. Tremblay, B.W. Smith, M.B. Leong, J.D. Winefordner: Spectrosc. Lett. **20**, 311 (1987)
7.80 J. Qu, Z. Zhou, L. Zhu, F. Lin: Appl. Phys. Lett. **59**, 271 (1991)
7.81 M.A. Cappelli, P.H. Paul, R.K. Hanson: Appl. Phys. Lett **56**, 1715 (1990)
7.82 M.J. Dyer, L.E. Jusinski, H. Helm, C.H. Becker: Appl. Surf. Sci. **52**, 151 (1991)
7.83 J.G. Liebeskind, R.K. Hanson, M.A. Cappelli: AIAA **91**, 2112 (1991)

Subject Index

ablation 11
- products 114
- threshold 31
- yield 45
absorbing inclusions 61
accumulated surface modification 76
accumulation effect 57
accumulation of defects 43
acoustic phonon scattering 59
adatoms 32
adiabatic potential energy surface (APES) 20
alkali halides 27
alkaline-earth fluorides 27
amorphous 29
angular rigidity 38
Auger 71
avalanche 59
- breakdown 41
- ionization 59
- ionization rates 60
- threshold 59

backing gas 169
bare surface 55
beam diagnostics 55
Beer's law 11
biomolecules 7, 135, 147
BK-7 80
blackbody radiation 62
bond-orbital theory 36
boron-doped Buckminsterfullerene 143
Buckyball 5
buffer gas 141

CdTe 32
cleavage 65
cleaved surfaces 62, 67
clusters 8, 135, 140
CO_2 136
coatings 55
collisionally activated dissociation (CAD) 140, 146
color centers 11
columnar growth 62

conditioning 70
cone formation 88
configuration coordinate 34
cooling transitions 24
cracking 65, 71
crystalline SiO_2 29
cumulative modification 70, 79
cyclotron 137

damaged surface regions 57
defect centers 63
defect-initiated ablation 47
defects 13
density 176
- of electronic excitation 13
- of excitation 27
desorption 54, 63
diagnostic 167
dielectric discontinuities 62
dimerization 41
double-pulse damage thresholds 75
double-pulse threshold 77

E' point defect 29
electric field 56
- enhancements 62
electro-optic materials 87
electron affinities (EA) 145
electron avalanche 59
electron-energy distributions 60
electron energy loss spectroscopy 71, 72
electron-hole plasma 15
electron-lattice interactions 13
electronic defects 11
electronic desorption 66
electronic excitation 107
electrons 63, 67
emission 63, 67
enhancement 56, 62
excited singlet 108
excitons 14, 70
external cluster source 141
extrinsic absorption 61

F-centers 79

Subject Index

ferro-electric materials 87
fluoride crystals 82
Fourier transform ion cyclotron resonance mass spectrometer (FTMS) 135
fractoemission 65, 71
Franck-Condon transition 33
free carrier absorption 58, 59, 79
free-electron heating 17
Frenkel pair 27
fullerenes 8, 143

GaAs 32
GaP 32, 37, 71
graphite 73
growth nodules 74
HfO_2 70, 79
high-T_c superconductors 5, 7, 85, 165
hot electron 60

incubation 109
infinite-layer superlattices 97
interband transition 72
interference 53
interferometry 172
intermediate excited state 21
intermediate states 68
inverse bremsstrahlung 61, 169
ion beam assisted deposition (IBAD) 98
ion detection 138
ion formation 136
ion microprobe analysis 4
ion structural techniques 139
ion manipulation 135
ion–molecule reactions 142, 144
ion trapping 135, 137
ionization 59, 136
ionization potentials (IP) 145
ions 63

Kapton^{TM} 110
kink sites 32
Knotek-Feibelman (KF) 36
Knudsen layer 163

Langmuir-Blodgett films 131
laser 136
 – ablation 12, 53, 108, 135, 157
 – desorption 135
 – – mass spectrometry 8
 – fusion 5
 – ionization mass analysis 71
 – mass spectroscopy 7
 – microprobe 3
 – sampling 157
 – surgery 7
laser-beam deflection 172

laser-induced desorption (LID) 63
laser–material interactions 157
lattice relaxation energy 21
LIBS 157
linear absorption 79
lithography 7
local optical electric field 62
localization 21
localization of two electrons or holes 35
localized excited state 21
low energy electron diffraction (LEED) 71
luminescence 21

MALDI-FTMS 152
mass spectra 158
materials processing technologies 11
matrix-assisted laser desorption/ionization or MALDI 147
Maxwellian distribution 69, 163
MBE 93
Menzel-Gomer-Redhead (MGR) model 33
MgO 29
micrograph 53
molecules 63
Monte Carlo calculation 46
morphology 57
MS^n 140
multi-hole localization 36
multilayer 70
multilayer optical coatings 73
multiphoton absorption (MPA) 58
multiphoton band-to-band transitions 15
multiphoton excitation 17, 63, 107
multiphoton photoelectric 67
multiphoton transitions 41
multiply charged ions 5

nanosecond pulsed lasers 11
nanotechnology 7
Nd:YAG 136
negative-U interaction 22
negative-U potential 35
neutral atoms 63
non-radiative transitions 15, 23, 24
non-stoichiometric regions 62
non-stoichiometry 71

oligonucleotides 148
optical breakdown 59
optical coatings 62, 70
optical phonon scattering 59
optical surface damage 53
organic photochemistry 107
oxide ferroelectrics 30
oxide glasses 29

Subject Index

particle deposition 89, 90
phase change 41
phonon kicking 24, 25
photo-induced fragmentation 140
photodissociation 142
photoemission 32
photokinetic etching 131
photothermal 80
photothermal deflection (PTD) 53, 68, 81
photothermal reflectivity (PTR) 81
picosecond laser pulses 77
plasma 12, 57
plasma–surface interactions 12
plasmon effects 6
platinum inclusions 61
plume diagnostics 6
PMMA 109
polaron 18
polished surfaces 62
polyimide 110
post excitation 6
post ionization 6, 160
pulse duration 55
pulsed-laser deposition (PLD) 85
pump-probe techniques 54

radiative 15
radiative recombination 24
recombination 20
refractive index 56
relaxation time 80
repetitive laser pulses 56
resonance effects 6
resonance ionization spectroscopy 30
rotational excitation 107

sapphire (Al_2O_3) 29, 68, 73
scanned-probe techniques 54
scanning force microscope 73
scanning force microscopy (SFM) 71
Schlieren photography 172
screened Coulomb potential 38
screened electron emission 71
seed electrons 61
self-trapped hole 22
self-trapped exciton (STE) 22
self-trapping 21, 33
shadowgraphy 172
shadowing effects 88
shifted Maxwellian velocity distribution 163
shock wave 167
SiO_2 70, 79
soft ionization 7
SQUID 95
STE 26
strategic defense initiative (SDI) 8

stress wave 116
strong-coupling solids 42
strong electron–lattice coupling 21, 33
sublimation 66
superlattices 95
supersonic expansion 142
surface analysis 71
surface electronic states 41
surface electronic structure 68
surface-induced dissociation (SID) 153
surface modification 57, 131
surface morphology 68
surface roughness 62
surface states 64, 67, 72

tandem-mass spectrometry 140
temperature 177
thermal diffusion 55
thermal diffusivity 80
thermal stresses 61
thermalization 19
thin-film 85, 165
Thomson scattering 175
three-dimensional electron-hole pairs 16
threshold fluence 111, 116
time-of-flight 70
transient grating 53
two-carrier localization 21
two-dimensional electron-hole pairs 16
two-hole localization 22, 35, 36, 38
two-hole localized state 22
two-photon absorption 78

ultrahigh mass resolution 139
ultrahigh vacuum 136
ultrathin films 96

vacancy clusters 45
velocity 179
– distributions 64
vibrational 107
volume explosion 122
volume plasmon 72
VUV 5

weak electron–lattice coupling 21, 35, 43
weak-coupling solids 44

X-ray emissions 5
X-ray lasers 8

YBCO films 87
yield-fluence relation 28

ZnO 30
ZnS 66, 77

Printing: Mercedesdruck, Berlin
Binding: Buchbinderei Lüderitz & Bauer, Berlin